DIRECTIO

Integrating Quantitative and Qualitative Research in Development Projects

Michael Bamberger
Editor

The World Bank
Washington, D.C.

© 2000 The International Bank for Reconstruction
and Development/THE WORLD BANK
1818 H Street, N.W.
Washington, D.C. 20433, USA

All rights reserved
Manufactured in the United States of America
First printing June 2000

The findings, interpretations, and conclusions expressed in this book are entirely those of the authors and should not be attributed in any manner to the World Bank, to its affiliated organizations, or to members of its Board of Executive Directors or the countries they represent. The World Bank does not guarantee the accuracy of the data included in this publication and accepts no responsibility for any consequence of their use. The boundaries, colors, denominations, and other information shown on any map in this volume do not imply on the part of the World Bank Group any judgment on the legal status of any territory or the endorsement or acceptance of such boundaries.

The material in this publication is copyrighted. The World Bank encourages dissemination of its work and normally will promptly grant permission to reproduce portions of the work.

Permission to *photocopy* items for internal or personal use, for the internal or personal use of specific clients, or for educational classroom use is granted by the World Bank, provided that the appropriate fee is paid directly to the Copyright Clearance Center, Inc., 222 Rosewood Drive, Danvers, MA 01923, USA; telephone 978-750-8400, fax 978-750-4470. Please contact the Copyright Clearance Center before photocopying items.

For permission to *reprint* individual articles or chapters, please fax a request with complete information to the Republication Department, Copyright Clearance Center, fax 978-750-4470.

All other queries on rights and licenses should be addressed to the Office of the Publisher, World Bank, at the address above or faxed to 202-522-2422.

Deborah Thornton prepared the first draft of this book based on transcripts of the workshop and provided editorial support for subsequent drafts.

Cover design by Grammarians; cover photograph by Curt Carnemark for the World Bank.

ISBN 0-8213-4431-5

Library of Congress Cataloging-in-Publication Data

Integrating quantitative and qualitative research
  / edited by Michael Bamberger.
      p.   cm. — (Directions in development)
  Includes bibliographical references.
  ISBN 0-8213-4431-5
  1. Economic development—Research. 2. Developing countries—
Economic conditions—Research. 3. Social sciences—Research.
  I. Bamberger, Michael.   II. Series: Directions in development
(Washington, D.C.)
  HD77 .I57   1999
  338.9'007'2—dc21   99–12980
                                                                                                    CIP

Contents

Preface..ix

PART I THE NEED FOR INTEGRATED RESEARCH APPROACHES IN SOCIAL AND ECONOMIC DEVELOPMENT..1

1. Opportunities and Challenges for Integrating Quantitative and Qualitative Research 3
   Michael Bamberger

   Evolution of World Bank Interest in Multimethod Approaches 4
   Defining Quantitative and Qualitative Methods 8
   Toward an Integrated Approach to Development Research 16
   Chapter Overviews 26
   Notes 32
   References 34

2. Issues and Approaches in the Use of Integrated Methods 37
   Kimberly Chung

   What Are Qualitative Data? 38
   Qualitative Data Collection Methods 40
   A Comment on Participatory Research 41
   Some Problems That Survey-Based Researchers Cite as Barriers to Accepting Qualitative Work 42
   Reasons to Integrate Qualitative and Quantitative Research Approaches 43
   Integrating Qualitative and Quantitative Methods 45
   Prerequisites for Conducting Integrated Research 45
   Notes 46

3. Gender Issues in the Use of Integrated Approaches 47
Roberta Spalter-Roth

Sociologists and Research Methodologies 47
Avoiding False Divides 48
Improving Quantitative Data for Studying Gender Issues 49
Giving Voice to Women 50
Integrated Approaches to Gender Research 50
Conclusion 52
Notes 52
References 53

PART II LESSONS FROM THE FIELD 55

Poverty Analysis
4. Integrated Approaches to Poverty Assessment in India 59
Valerie Kozel and Barbara Parker

Research Objectives 59
Approaches and Assumptions 60
Implementation of the Research 61
Selection of Research Sites 63
Preliminary Findings 63
Lessons Learned 66
Conclusions and Recommendations for Future Research 66
Notes 68
References 68

5. Studying Interhousehold Transfers and Survival Strategies of the Poor in Cartagena, Colombia 69
Gwyn Wansbrough, Debra Jones, and Christina Kappaz

Background on the Study and Research Methodology 69
Quantitative Research 71
Qualitative Research 71
Timing and Costs 72
Issues and Observations Regarding Integrated Methodologies 73
Preliminary Research Findings 77
Gender and Transfer Giving and Receiving 79
Notes 81
References 81

Education
6. Evaluating Nicaragua's School-Based Management Reform 85
Laura Rawlings

Purpose and Background of the Evaluation 85
The Reform Program 86
Research Sources and Methods 87
Findings of the Evaluation 90
Impact of the Evaluation 94
Stage II Research 94
Value of the Mixed-Method Approach 95
Notes 96
References 96

7. Evaluating the Impacts of Decentralization and Community Participation on Educational Quality and the Participation of Girls in Pakistan 98
Guilherme Sedlacek and Pamela Hunte

Research Objective 98
Types of Studies Conducted 98
Qualitative Research Design 99
Sampling Methods 100
Research Questions 100
Findings and Implications 100
Follow-On Research 102
Recommendations Regarding Integrated Research 102

Health, Water Supply, and Sanitation
8. Evaluating the Impact of Water Supply Projects in Indonesia 107
Gillian Brown

Overview of the Research 107
Overview of the Projects 107
Research Methodology 108
Data Analysis 109
Phase 1 Findings 110
Gender Analysis 111
Advantages of Integrated Research 112
Disadvantages of Integrated Research 112
Recommendations for Future Research 112

9. Social Assessment of the Uzbekistan Water Supply, Sanitation, and Health Project 114
Ayse Kudat

Background 114
The Research and Consultation Elements of Social Assessment 115
Research Methodologies 115
Establishing a Social Science Network 117
Needs Assessment 117
Household Study 118
Preliminary Findings 118
Subsequent Social Research 119
Communicating the Findings 119
Impact of the Social Assessments 119
Notes 120
References 120

10. Using Qualitative Methods to Strengthen Economic Analysis: Household Decision-Making on Malaria Prevention in Ethiopia 121
Julian Lampietti

Motivation 121
Rapid Ethnographic Study 123
Household Survey 123
Validity and Reliability of Quantitative Survey Results 125
Qualitative and Quantitative Perceptions of Malaria 127
Responses to Valuation Questions 128
Conclusions 130
Notes 131

Women and Children
11. UNICEF'S Use of Multiple Methodologies: An Operational Context 135
Mahesh Patel

Introduction 135
Monitoring 136
Evaluation 137
Research 139
Conclusion 140
Notes 141
References 142

Part III Lessons Learned 143

12. Lessons Learned and Guidelines for the Use of Integrated Approaches 145
Michael Bamberger

The Benefits of Integrated Research 145
Operational Approaches to the Integration of
Quantitative and Qualitative Methods at Each Stage
of the Research Process 149
Integrated Approaches during Formulation of
Research Questions 150
Integrated Approaches during Research Design 151
Integrated Approaches during Data Collection 151
Integrated Approaches during Data Analysis and Interpretation ..151
Conducting Research on Different Levels 152
Modalities of Integration 154
Operational Implications 155
Challenges in Using Integrated Approaches 157
Notes 164
References 164

Bibliography 165

Tables
1.1 Characteristics of Quantitative and Qualitative Approaches to Sample Selection, the Research Protocol, Data Collection, and Data Analysis 11
1.2 Elements of an Integrated, Multidisciplinary Research Approach ..25
5.1. Allocation of Time by Task in the Cartagena Study 73
5.2. Proportion of Households Giving or Receiving Transfers by Income in Quintiles 78
5.3. Proportion of Households Giving or Receiving Transfers by Gender of Head of Household 80
6.1 Nicaragua's School Autonomy Reform: Changes in the Locus of Decision-Making Over Time 88
10.1 Comparison of Qualitative and Quantitative Welfare Rankings ..126
10.2 Perception of Seriousness of Malaria 128
10.3 Number of Hypothetical Vaccines Purchased by Price 129
12.1 Frequently Cited Strengths and Weaknesses of Quantitative Research Methods 146
12.2 Frequently Cited Strengths and Weaknesses of Qualitative Research Methods 147

12.5 Elements of an Integrated, Multidisciplinary Research
   Approach 153
12.3 Quantitative and Qualitative Methods Used in the Case Studies
   Presented in this Report 159
12.4 Examples of the Integration of Quantitative and Qualitative
   Methods in the Case Studies Presented in this Report 162

Figures
10.1 Comparison of Qualitative and Quantitative Questions on Food
   Production 126
10.2. Malaria Vaccine Demand for 200-Household Village in Tigray,
   Ethiopia 130

Preface

This report is based on a two-day workshop held in June 1998 at which outside research specialists and World Bank staff discussed the importance of integrating quantitative and qualitative research methods and reviewed experiences in the use of mixed-method approaches in Bank research and project design. The report traces the growing interest of the World Bank in the integration of two traditionally distinct research traditions: quantitative survey research and participatory and other qualitative research methods.

Researchers have recognized over the past few years that quantitative analysis of the incidence and trends in poverty, while essential for national economic development planning, must be complemented by qualitative methods that help planners and managers understand the cultural, social, political, and institutional context within which projects are designed and implemented. This is reflected in the recent report, *Poverty Reduction and the World Bank: Progress in Fiscal 1998* (World Bank 1999), which devotes a chapter to the discussion of the integration of *contextual* and *noncontextual* methods in data collection and analysis and the increasing use of participatory poverty assessments. Interest in the use of integrated quantitative and qualitative methods is also growing in areas such as resettlement, dialogue with nongovernmental organizations (NGOs) and other sectors of civil society, social development and participatory assessment, and the use of social assessment in project design and policy formulation.

Part I, *The Need for Integrated Research Approaches in Social and Economic Development*, contains three chapters describing the evolution of interest in and the potential benefits of integrated research. The main qualitative methods are described and the importance of integrated approaches in gender analysis are discussed.

Part II, *Lessons from the Field*, contains seven chapters presenting case studies on how integrated approaches have been used in poverty analysis (chapters 4 and 5), education (chapters 6 and 7), and health, water supply, and sanitation (chapters 8 to 10). Chapter 11 describes the use of mixed-method approaches by UNICEF.

The final chapter (chapter 12) discusses lessons learned with respect to when and how to use integrated approaches at different stages of the project cycle and assesses the benefits that can be achieved. This chapter also includes reference tables identifying the quantitative and qualitative methods discussed in each chapter and the strategies used to ensure effective integration.

Part I
The Need for Integrated Research Approaches in Social and Economic Development

1
Opportunities and Challenges for Integrating Quantitative and Qualitative Research

Michael Bamberger[1]

There is a growing recognition of the benefits to be gained from combining quantitative and qualitative methods in development research. A number of areas are identified in which the World Bank, in common with other development agencies, is making increasing use of integrated quantitative and qualitative research approaches. There is no clear distinction between quantitative and qualitative methods, and it is more helpful to consider data collection and analysis methods as being located on a quantitative-qualitative continuum. Many research designs use a combination of quantitative and qualitative methods at different stages of the research cycle. The major characteristics of quantitative and qualitative approaches are discussed for each stage of the research process, and are illustrated with examples drawn from the case studies presented in chapters 4-11. An innovative approach in recent World Bank poverty assessments combines quantitative and qualitative data with contextual and noncontextual data collection and analysis methods. The chapter concludes with guidelines for developing an integrated research methodology that ensures that full integration of quantitative and qualitative methods is achieved in the analytical framework and at all stages of the research process.

The desirability of integrating quantitative and qualitative research methods in development work is widely acknowledged, but the successful implementation of integrated approaches in the field has often proved elusive. However, there is now a growing body of experience in the development field, some of which is reported in this volume, demonstrating the benefits to be achieved from multimethod research integrating quantitative and qualitative methods.[2]

However, despite significant progress in promoting integrated approaches, many researchers from both quantitative and qualitative traditions still often find it difficult to make full use of the data collection methods and analysis from the other tradition. Some **quantitative survey researchers** may find it difficult to make full use of the wealth of case studies, PRA maps, calendars, and key informant interviews they have commissioned. Quantitative researchers also complain that their important messages on the incidence and determinants of key development variables such as malnutrition, usage of health services, and consumption-expenditure measures of poverty, and on the poverty consequences of economic variables such as price changes, agricultural marketing policies, and so on are dismissed as "too macro" by many qualitative researchers. On the other hand, *qualitative researchers* often complain that their findings may be dismissed by survey researchers as not being sufficiently representative or rigorous. Qualitative researchers also express the concern that even after collaborative research efforts, the survey researchers still do not understand the true nature of a complex phenomenon such as poverty.

The message is that there is a growing consensus on the value of integrated approaches, but that further work is needed to develop guidelines for the effective use of these integrated approaches. Most of this publication is devoted to presenting examples of promising approaches to integrated quantitative and qualitative research which have been used in the World Bank and other development institutions. The purpose of this chapter is to provide a framework for assessing these experiences and for discussing the lessons that can be learned. After tracing the increasing interest of the World Bank and other development agencies in the use of integrated research approaches, this chapter addresses the following four issues: (a) the definition of qualitative and quantitative research methods; (b) potential benefits from the use of integrated approaches; (c) issues and challenges in the use of integrated approaches in development work; and, finally, (d) whether there is an integrated research approach that consists of more than simply using a wider range of data collection methods.

Evolution of World Bank Interest in Multimethod Approaches

A number of factors have contributed to the World Bank's growing interest in the integration of quantitative and qualitative research methods. The first is the Bank's focus on poverty. Much of the early work on poverty was highly quantitative: how many people fall below the poverty line, how this number varies during different kinds of economic change, and so on. It became increasingly clear, however, that while numbers are essential for policy and monitoring purposes, it is also important to understand people's perception of poverty and their mechanisms for coping with poverty and other situations of extreme economic and social stress. Kozel and Parker's study on poverty in northern India (chapter 4) illustrates the wide range of coping mechanisms and different attitudes to the possibility of escaping from poverty. Although many poverty alleviation programs are based on providing opportunities for economic improvement within the community where people live, their study showed that many families believe they have no way to escape from poverty, while others see migration as the only possible escape.

Poverty Reduction and the World Bank: Progress in Fiscal 1998 summarizes recent developments in the use of mixed-method approaches in the Bank's work on poverty analysis.[3] While almost all World Bank poverty assessments (PAs) rely on data collected through sample surveys, an increasing number combine this with data collected through participatory poverty assessment (PPA) techniques and analyses carried out with and by poor people in the field. The report states that while earlier PPAs contributed mainly to enriching the *description* of poverty and seldom followed an integrated approach, more recent work shows three important changes: greater integration of methods, more emphasis on understanding the *causes* of poverty, and greater participation.[4]

The report distinguishes between data, which can be either quantitative or qualitative, and data collection and analysis methods, which can be either contextual or noncontextual. "Contextual methods attempt to understand human behavior within the social, cultural, economic, and political environment of a locality, usually a village or neighborhood or social group." Noncontextual methods, on the other hand, "abstract from the particularities of a locality to gauge general trends."[5] Recent poverty assessments seek to enrich the understanding of poverty through the integration of contextual and noncontextual methods. The report argues that it is helpful to think of two continua: one where methods used are more or less contextual, and the other where data gathered are more or less quantitative.

Hentschel (1998, 1999) illustrates the use of contextuality to characterize information needs for planning the utilization of public health services. He argues that in the public health sector, and in the social sciences more generally, rather than considering quantitative and qualitative methods as describing two different realities, they are both needed to describe and understand one reality. He also argues that "labeling both methods and data as quantitative or qualitative creates a problem with regard to analyzing what the comparative advantages of different methods and data types are to understand human behavior like the utilization of health facilities." Hentschel's paper illustrates three ways in which contextual and noncontextual methods can be combined:

- Certain information can be obtained through contextual methods of data generation only. In these instances, strict statistical representability will have to give way to inductive conclusion, and to internal validity and replicability of results.

- In many instances, contextual methods are needed to design appropriate noncontextual data collection tools.
- If information requires noncontextual data collection methods, contextual ones can nevertheless play an important role for assessing the validity of the results at the local level.[6]

A number of other recent studies report progress in developing subjective methods of poverty assessment, many of which are specifically designed to be used in parallel with conventional survey approaches. Pradhan and Ravallion present recent findings on the use of subjective poverty lines.[7] They report that, on average, these accord closely with "objective" poverty lines, although there are notable differences when regional and demographic profiles are constructed. Ravallion and Lokshin report that current household income relative to a poverty line can only partially determine how Russian adults perceive their economic welfare.[8] Other factors such as past incomes, individual incomes, household consumption, current unemployment, risk of unemployment, health status, education, and relative income in the area of residence all influence perceptions of welfare.

Mangahas describes a different approach to poverty assessment adopted by the Social Weather Stations program in the Philippines,[9] which has used a small set of qualitative measures to monitor changes in the level of poverty in the Philippines for over 20 years.[10] These self-rating indicators consistently estimate a higher proportion of households experiencing poverty than do the official poverty indicators. In addition to the economy and speed with which these measures can be applied and analyzed, Mangahas argues that these indicators provide a better reflection of the impact of short-term economic and political changes on poor and vulnerable groups. Somewhat similar methodologies are used, although for different purposes, by the Eurobarometer's "gainers/losers" and "optimists/pessimists" indicators,[11] and the U.S. Conference Board's consumer confidence index.

The Bank's work on resettlement is another area where there has been a great deal of interest in integrated research methods. The Bank has had a strong focus on understanding the processes of resettlement, which traditionally has been an area for anthropological research. However, there is a need for quantitative data when it comes to evaluating the impact of a large-scale resettlement program that involves the relocation of tens of thousands of people.

As the Bank has become more involved in dialogue with civil society—including nongovernmental organizations (NGOs), women's organizations, and academic groups—it has become apparent that different groups employ different paradigms for addressing issues such as structural adjustment, gender, or household structures. As a result, there have been some major efforts to develop a common framework for assessing whether people are better or worse off as a result of these economic reforms. One example is the Structural Adjustment Participatory Research Initiative (SAPRI), in which NGOs and Bank researchers are trying to come up with a broader, more integrated approach for evaluating the impacts of structural adjustment. That is still very much an ongoing dialogue, but there has been some important progress.

As the Bank has moved into the field of social development, there has been a greater focus on participatory assessment methods and other qualitative approaches, such as rapid rural appraisals. This has given rise to a number of methodological questions regarding how to present qualitative research findings so as to increase their legitimacy in the eyes of quantitatively oriented policymakers and planners who wish to know whether the findings of specific cases can be generalized to wider populations. For example, how can you determine whether the findings of a focus group represent the views of the community, and not just the position of an influential individual or minority?

Another promising area for the use of a multimethod approach is the Bank's social assessment work, in which researchers are using a variety of quantitative and qualitative tools to plan and monitor development interventions. The objective of these studies is to understand the processes of change, determine the number of people affected, and identify socially and politically viable options for interventions, such as introducing new transport systems, privatizing industries, improving the economic viability of water supply systems, and so on. Chapter 9 illustrates how integrated approaches were applied in a recent social assessment study for a proposed water supply, sanitation, and health project in Uzbekistan.[12]

Incorporating quantitative methods of data collection and analysis into previously qualitative research areas is becoming important, both in the Bank's operational work and to increase the legitimacy of qualitative data. For example, even when they themselves agree with the findings of the qualitative studies, policymakers often want the qualitative findings to be presented using conventional statistical principles, in order to make these findings more acceptable to other agencies. The need to increase the reliability of qualitative findings through the use of appropriate statistical techniques is often important to ensure the acceptability and utilization of findings

There has been significant interest in developing rapid, cost-effective assessment methods to provide timely feedback on the likely social costs and outcomes of proposed projects and policies. There may only be a few weeks to conduct such a study, either for administrative reasons (for example, because loan documents must be prepared by a certain date) or because decisions must be made in the midst of a crisis. Consequently, there is a strong desire to find a way to do things quickly, but with sufficient rigor that the findings can be defended and generalized. The recent financial crises in Southeast Asia stimulated interest in this application of rapid social assessments because the crises, and the measures taken to address them, often had serious repercussions for poor or vulnerable groups.

The need for a different research paradigm has also been identified in the area of gender analysis. Conventional survey methods are often inadequate for capturing the views of women in male-dominated societies where only the men provide information to outsiders, or where the women feel intimidated about responding freely to research questions. The growing interest in participatory research has emphasized the fact that the views of women are frequently not captured unless special gender-sensitive methods are devised to give them voice (Bamberger, Blackden, and Taddese 1994). There has been a great deal of discussion about ways to integrate economic analysis with gender analysis in order to adequately give voice to women and to understand how development affects women and men. An important area in which innovative research methods have been conducted concerns the assessment of women's time poverty through the measurement of the hours per day which women spend on each of their multiple tasks (Blackden and Morris-Hughes 1993). Time poverty analysis and the development of methods to obtain more realistic estimates of the value of women's time is proving to be a key issue in making transport projects more gender sensitive (Barwell 1996; Bryceson and Howe 1993; Bamberger and Lebo 1999). Some of these issues are discussed in chapter 3.

The demand for integrated research approaches comes from researchers with a quantitative background (for example, economists working on poverty assessments) as well as from those who have traditionally used more qualitative field studies (for example, work on resettlement and social assessment). The case studies presented in chapters 4-11 show that integrated approaches appeal in different ways to researchers coming from different traditions.

Defining Quantitative and Qualitative Methods

Most attempts to draw clear distinctions between quantitative and qualitative research have floundered in the face of the many counter examples that can be found to challenge each categorization. Chung (chapter 2) cites a number of caveats for every simple definition. For example, while textual reports on key informant interviews and group discussions will often be analyzed using broad qualitative interpretations, it is also possible to conduct the analysis using rigorous statistical content analysis. On the other hand, the design of the most highly structured questionnaires usually involves subjective, qualitative decisions on the selection of topics to be included and the way that questions are worded.

This section discusses the distinction between quantitative and qualitative methods in different stages of the research process. The next section addresses the question of whether it is possible to identify an integrated approach to, or a unique comparative method for, development research that consists of more than simply broadening the range of data collection and analysis methods.

When attempting to define the differences between quantitative and qualitative research methods, it is useful to think of methods of sample selection, design of the research protocol, data collection and recording, and data analysis as each being ranged along a quantitative/qualitative continuum. While some studies rely exclusively on quantitative methods for sampling, data collection, and data analysis, and others rely exclusively on qualitative methods, many studies mix and match statistical sampling techniques, qualitative data collection, and statistical analysis from the qualitative and quantitative traditions. It then becomes an empirical question as to whether a particular discipline (for example, demography or economics) tends to rely more heavily on tools and methods from the quantitative end of the continuum than does another discipline or subdiscipline (such as empowerment evaluation or participatory research). Even when a particular discipline has a general penchant for either quantitative or qualitative methods, many exceptions will be found.[13]

Hentschel's (1999) distinction, discussed earlier in this chapter, between contextual and noncontextual methods of data collection and data analysis is also helpful in this context. Contextual methods attempt to understand human behavior within the social, cultural, economic, and political environment of a locality, usually a village or neighborhood or social group; while noncontextual methods abstract from the particularities of a locality to gauge general trends.

It is also helpful to distinguish between the two main purposes for qualitative research. The first is *exploratory research*, in which the objective is to understand the context within which behavior is determined or processes take place in order to develop hypotheses, or to present case studies of particular communities, groups, or individuals. In these studies methods can be flexible and unstructured, with little or no concern for comparative rigor. The second purpose is *directed* research, in which hypotheses are tested, or rigorous comparative findings are required. For these latter kinds of studies, both qualitative and quantitative researchers use similar sampling methods, follow standard measurement and reporting procedures, and observe careful documentation requirements.

Table 1.1 describes typical quantitative and qualitative approaches to sample selection, design of the research protocol, data collection techniques, and data analysis. The primary characteristics of the two approaches in each of these areas are described in more detail below.

Procedures for Selection of Subjects

An essential characteristic of *quantitative* research is random selection of subjects so that each subject has an equal or known probability of selection. This makes it possible to generalize from the sample to the total population. Equally important, the selection procedures make it possible to determine the statistical significance of differences between subgroups.

Qualitative research, by comparison, has no single defining sampling procedure; rather, the choice of sampling method is determined by the purpose of the study. Miles and Huberman talk about the "bounding function" of the conceptual framework in defining sample selection procedures.[14] In exploratory studies, or where the purpose of the study is to obtain a general understanding of the attitudes or priority concerns of a community, there may be no clearly defined selection procedure; any interested members of the community may be invited to participate in focus groups. One of the most common methods is purposive or theoretical sampling in which interviews are conducted with representatives of each category, stakeholder, or socioeconomic group of interest to the objective of the study, but without random selection of the particular subjects who are studied in each group. Kozel and Parker use this approach in their study of poverty in India presented in chapter 4.

There are, however, many examples in which random sampling procedures are used to select the subjects for qualitative interviews. A promising and practical approach to ensuring a certain degree of representativity is to draw the sample of communities or individuals for the qualitative research from the sampling frame that is used for the quantitative stage of the research (see chapters 4, 6, and 8). Qualitative research may also be interested in the selection and analysis of outliers in order to explain the reasons for the deviation from the general pattern observed in the data. In the Indonesia water supply study (chapter 8), this approach is used to follow up on the one community where women were not involved in water supply management.

Research Protocol

Another difference between qualitative and quantitative work is found in the area of research protocol. Whereas in quantitative research it is extremely important to administer a survey according to a standard pro-

Table 1.1 Characteristics of Quantitative and Qualitative Approaches to Sample Selection, the Research Protocol, Data Collection, and Data Analysis

Research Activity	Quantitative Approach	Qualitative Approach
Selection of subjects or units of analysis	• Random sampling to ensure findings can be generalized, and to permit statistical testing of differences between groups • Selection methods clearly documented	• Choice of selection procedure varies according to the purpose of the study (exploratory or directed). • Purposive or theoretical sampling often used to ensure representation of all important groups • Representativity can be ensured by selecting cases as a subsample of a quantitative sample survey. • Random sampling methods can be used
Research protocol	• Data usually recorded in structured questionnaires • Extensive use of precoded, closed-ended questions • Standard protocol must be followed consistently throughout the study	• Protocol may be unstructured with information being entered in the form of narrative text • In some studies the protocol may be modified during the course of the study.
Data collection and recording methods	• Mainly numerical values (integer variables) or closed-end (ordinal or nominal) variables which can be subjected to statistical analysis. • Some open-ended questions may be included. • Observational checklists with clearly defined categories may be used.	• Textual data: sometimes recorded verbatim, sometimes in notes • Informal or semistructured interviews • Focus groups and community meetings • Direct observation • Participatory methods • Photographs • Sociometric charts • Behavior or unstructured interviews may be recorded into precisely defined categories. Laptop computer templates may be used for this purpose.

(Table continues on the following page.)

Table 1.1 (*continued*)

Research Activity	Quantitative Approach	Qualitative Approach
Triangulation	• Consistency checks are built into questionnaires to provide independent estimates of key variables. • Qualitative methods (for example direct observation) used to check responses to questions.	• Several qualitative methods used for consistency. • Monitors participate in focus-groups, etc., to provide an independent assessment of the findings. • Studies often coordinated with sample surveys so that each can provide a check on the other.
Data analysis	• Descriptive statistics (indicators of dispersion and central tendency) • Multivariate analysis to examine factors contributing to the magnitude and direction change. • Significance tests for differences between groups	• Each subject treated separately (e.g., case studies) to examine the unique characteristics of each person or group; or • Numerical analysis to permit systematic comparison of individuals, communities or groups. • Analysis emphasizes context of study and how it affects understanding of findings • Follow-up to statistical analysis of quantitative surveys to examine statistical outliers, or to put "flesh and bones" on the statistics by preparing case studies on main categories studied.
Role of the conceptual framework	• The conceptual framework leads to the formulation of hypotheses that can be empirically tested.	• The conceptual framework may lead to the formulation and testing of hypotheses. • When the purpose is to explore the uniqueness of each situation, the conceptual framework may be developed through a process of iteration. Subjects may be observed over a period of time, and the framework may be continuously revised on the basis of new information.

The quantitative framework can often, but not always, be characterized as:
- Starting from the macro, rather than the micro level
- Focused on outcomes rather than processes
- Positivist

The qualitative framework can often, but not always, be characterized as:
- Starting at the individual level and seeking to understand the constraints of everyday life.
- Seeking to understand processes as well as outcomes
- Frequently (but not always) using a subjectivist approach to understand how the world is perceived by particular individuals or communities.
- Holistic: putting subjects into their socioeconomic context

tocol, many qualitative research protocols are relatively unstructured and flexible. When the purpose of the study is to understand the unique characteristics of a particular community, organization, or group, qualitative researchers will often modify the format or content of a study while it is in progress in order to capitalize on what is learned in the field and to pursue certain preliminary findings. However, as indicated above, when the purpose is to ensure comparability, a precisely defined research protocol will also be used for qualitative research.

Data Collection and Recording Methods

In *quantitative* research, information is usually collected and recorded either numerically or in the form of precoded categories. The principal advantage of such surveys is that they can be administered to large numbers of individuals, organizations, or households using standardized methods.

In *qualitative* studies, information is most frequently recorded in the form of descriptive textual reports with little or no categorization. The documentation may consist of subjects' responses to semistructured interview questions, notes taken during focus groups, or other kinds of group interaction, or the researcher's observations of relevant aspects of a community or organization. In other cases the information may be recorded within predefined categories but with reports presented in an unstructured or semistructured form within each category.

It should be noted, however, that some qualitative studies do rely on precoded classification of data, such as the perceived level of understanding of campaign communications, level of formality of community groups, and so on. The Indonesia water supply study (chapter 8) uses this approach to classify the effectiveness of community organization in project implementation, while the Nicaragua and Pakistan education studies (chapters 6 and 7, respectively) use it to evaluate different aspects of the quality of education in those countries.

Triangulation

Triangulation is the principle of increasing the validity of the data by looking at different data sources or by going back to the same subjects at different periods of time and asking the same kinds of questions. The purpose of triangulation is to improve the validity of one's findings.

Some researchers claim that triangulation is a characteristic of qualitative research. However, triangulation is also used in quantitative research, usually by comparing findings from different surveys, or by comparing survey findings with census data.

One of the most important ways in which triangulation can be used is to compare the findings of qualitative and quantitative studies. The Cartagena study (chapter 4) uses participant observation and informal interviews to double check the sources and volume of transfers reported in the sample survey.

Data Analysis

Quantitative methods for data analysis are most commonly analyzed using descriptive statistical methods such as measures of dispersion or central tendency or multivariate analysis to examine the factors contributing to the direction and magnitude of change. In addition, statistical tests will often be used to test the significance of differences between groups or of changes over time. Standardization across observations makes it possible to aggregate measures and to make statistical comparisons among individuals, households, regions, and time periods.

Data collected using qualitative methods can be either analyzed and interpreted descriptively or numerically. The objective of descriptive analysis is to understand the unique characteristics of a particular context (i.e., community, organization, event), household, or individual. Numerical analysis permits systematic comparison of these entities. A wide range of analytical methods are available for converting textual and image data (photographs, advertisements, etc.) to a numerical format.

Qualitative research is frequently more interested in eliciting the stories behind particular individuals or groups. In some cases the emphasis will be on statistical outliers or unusual cases that are not behaving as expected, whereas in other cases the purpose is to put "flesh and bones" on the findings of the statistical analysis. For example, a researcher conducting a nutritional study might notice that in a high-income household, where one would expect the children to be well nourished, they are instead quite thin and stunted. This finding might lead the researcher to do a case study of that household to try to uncover the dynamics behind that unexpected outcome.

In other instances, case studies may be prepared on typical households or communities to help understand the meaning of the statistical findings. In an earlier study on interhousehold transfers on which the research described in chapter 4 was based, the statistical analysis identified a number of interhousehold transfer strategies, including transfer avoiders, transfer givers, transfer receivers, and households seeking to develop transfer networks.[15] Case studies were then prepared to illustrate each of these strategies.

Conceptual Framework

Qualitative and quantitative research frequently differ in terms of the role of the conceptual framework. In quantitative research, the researcher's conceptual framework usually leads to the formulation of hypotheses, which are then tested. This can also be true for qualitative research, but when the purpose is to explore the uniqueness of each situation, the conceptual framework and research protocols may evolve in the field as data are obtained. This occurs through the process of iteration, the purpose of which is to clarify or follow up on information that was obtained in an earlier stage of the research.

While quantitative research ensures comparability over time by applying standard questions to the same or statistically comparable samples of households or individuals at different points in time, iteration in qualitative research does not always involve repeated contact with the same or comparable respondents or groups. In a qualitative study, the purpose may be to react to, and build upon, what was learned in the previous round. In some cases this may require that different kinds of respondents be added as the study progresses.

While there are many exceptions, qualitative conceptual frameworks can often be characterized as having a micro rather than a macro focus, seeking to understand processes starting at the individual rather than the aggregate level, and having a holistic focus. The framework and analysis also tend to be interpretive, often relying on naturalistic observation to capture the constraints of everyday life.[16]

Toward an Integrated Approach to Development Research

Although some writers are concerned that the overenthusiastic adoption of qualitative methods may negatively affect the quality of quantitative research,[17] it is now widely acknowledged that there are considerable benefits to be gained from combining quantitative and qualitative methods. Kertzer and Fricke (1997) state that there is a " ... growing recognition of the limitations of the survey mode data collection for gathering accurate, fully textured, and nuanced data at multiple levels of social reality"[18] and emphasize that surveys must be complemented by qualitative methods.

Despite increasing eclecticism in the combination of data collection methods, there is much less integration at the level of the conceptual framework and the overall research approach. A demographer or economist may use focus groups or observational techniques to enrich survey data, or a social anthropologist might include a rapid sample survey to compare the socioeconomic characteristics of the case study households with those of the community. In general, however, most researchers have found it difficult to break out of the conceptual framework of their own discipline.

Although quantitatively oriented researchers such as demographers and economists make increasing use of qualitative data collection methods, Obermeyer (1997) argues that it has proved much harder for them to accept the analytical frameworks used in the disciplines from which qualitative data collection methods are borrowed.[19] Some have argued that the popularity of focus groups is due to the fact that they are much less tied to particular analytical frameworks than are some other qualitative methods. Similarly, some economists feel that many qualitatively oriented sociologists and anthropologists are unwilling to give up their skepticism about sample survey research findings on subjects such as poverty.

A number of writers have argued that the greatest potential benefit from cross-disciplinary research would come from developing new integrated analytical frameworks where two disciplines could each draw on the conceptual and analytical frameworks of the other.[20] Rao, for example, would like to see ethnographic analysis being used to inform the development of rational choice models. While progress has been made, the potential for sharing disciplinary frameworks remains largely untapped.[21]

Ragin argues in favor of a distinct *comparative* research approach that is more than the simple integration of quantitative and qualitative methods. While qualitative researchers tend to consider multiple cases as many instances of the same thing,

> comparative researchers who study diversity, by contrast, tend to look for differences among their cases. Comparative researchers examine patterns of similarities and differences across cases and try to come to terms with their diversity. Quantitative researchers . . . also examine differences among cases but with a different emphasis. In quantitative research, the goal is to explain the covariation of one variable with another, usually across many, many cases. Furthermore, the quantitative researcher typically has only broad familiarity with the cases included in a study. . . . The emphases of comparative research on diversity (especially the different patterns that may exist within a specific set of cases) and on familiarity with each case, makes this approach especially well suited for the goals of exploring diversity, interpreting cultural or historical significance, and advancing theory.[22]

Datta also stresses the important role of case studies as an integrating tool.[23]

A fully integrated research *approach* would draw on the conceptual and analytical frameworks of at least two disciplines in the design, analysis, and interpretation of the research, while at the same time combining a broad range of data collection methods—in most cases including both quantitative and qualitative methods. Although the integrated approach must be adapted to the needs of each specific study, a fully integrated research approach will normally seek to ensure integration at the following stages of the research process:

- Conceptual and analytical framework
- Exploratory analysis
- Sample selection
- Data collection methods
- Understanding the context of the research study
- Triangulation
- Data analysis and follow-up fieldwork
- Presentation of findings

Conceptual and Analytical Framework

An integrated approach can broaden the conceptual and analytical framework of a study. In many demographic or economic research projects, for example, the study is designed to test hypotheses concerning quantitative relationships among the variables in the model. Frequently, noncontextual data collection and analysis methods are used, and no contextual variables are included to take into consideration the unique characteristics of the social, economic, political, and cultural context within which the study is conducted. In these cases an integrated approach could strengthen the analysis by taking into consideration the influence of contextual variables such as social organization, culture or the political context.

In demographic research in particular, there is increasing interest in the effect of culture on demographic outcomes, but conventional quantitative research approaches have not been able to capture the subtleties and complexities of culture. Culture is either ignored in the modeling, or it is reduced to one or more dummy variables in what Obermeyer (1997) describes as the "add fieldwork and stir" approach.[24] Many demographers, and some economists, believe that one of the greatest potential contributions of anthropology to their discipline is to permit a fuller understanding of culture to be built into their models.[25] The analysis of culture requires that the conventional demographic or economic analytical framework be broadened to incorporate cultural concepts and methods of data collection and interpretation into the research design—it is not just a question of introducing a new data collection instrument.

Kertzer and Fricke discuss several factors which may constrain the full adoption of the anthropological approach to culture by demographers; the second and third of these points may also be relevant to economics and other quantitatively oriented disciplines.[26] First, most demographers are trying to adopt a structuralist-functionalist model of culture, which most anthropologists would consider to be at least 30 years out of date. Second, many demographers who are unfamiliar with anthropological theories and methods have tried to adopt certain concepts concerning culture without understanding the theories underlying the concept. Such borrowing out of context tends to greatly limit the utility of the ideas that are borrowed. A related point is that demographers have usually not wished to become involved with the ideological debates surrounding modern anthropology, some of which concern the interpretation of culture.

On the other hand, sociological, political science, and anthropological frameworks can also be incorporated into the design of economic research in areas such as poverty assessment (see the India poverty study described in chapter 4), social capital, interhousehold transfers (see the Cartagena income transfer study in chapter 5), and the evaluation of the impacts of many kinds of development interventions (see the evaluation of education projects in chapters 6 and 7, and the Indonesia water supply project in chapter 8). In each case the analytical framework and the whole research approach would have to be broadened to incorporate these new concepts.

Similarly, many kinds of anthropological, sociological, and political research can benefit from the incorporation of economic approaches. Rao argues that economics can play an important role by offering a number of analytical approaches—such as rational choice models and game theory—which can be applied to the kinds of rich observation of human behavior provided by anthropology and sociology and political science. He states:

> Economics is in the business of developing formal models and has spent countless person years constructing a quantitative discourse where models of human behavior are checked against survey data with statistical tools. These models assume some kinds of intelligent action – which does not have to be merely economically rational.[27]

Some of the recent World Bank poverty assessments are developing an integrated conceptual framework which combines contextual and noncontextual methods to explain how factors such as culture, community organization, and the local economy can explain variations in the ways that families with similar conditions in terms of noncontextual poverty and welfare indicators perceive their situation and their prospects for improvement (World Bank 1999).

Lampietti (chapter 10) illustrates how the economic concept of contingent valuation was combined with anthropological studies of household and community knowledge and attitudes about malaria to provide new insights into household decision-making with respect to the purchase of a hypothetical malaria treatment in northern Ethiopia.

Exploratory Analysis

Qualitative methods can be used to conduct exploratory analysis during the preliminary stages of a survey to understand the social, cultural, and political context affecting the communities to be covered by the study. These methods can also help with hypothesis testing and the definition of key concepts such as "the household," "work," and "vulnerability." The full value of this exploratory analysis will only be obtained if it is conducted early in the research design, and if a sociological, anthropological, or political science framework has been built into the conceptual framework. In the India poverty study (chapter 4), the exploratory analysis highlighted the importance of the caste system as a constraint on household perception of the possibility of escaping from poverty, and it showed that many families believed that leaving the village was the only possible way to escape.

Sample Selection

The use of statistical sampling procedures can help to ensure that the results of case study research and other qualitative methods can be generalized, thereby increasing the likelihood that the findings will be accepted and used by quantitatively oriented policy makers and planners (see chapters 4, 5, 6, 7, and 8 for alternative ways in which researchers tried to ensure the generalizability and credibility of their findings). On the other hand, exploratory research methods can be used to determine the criteria for the formulation of cluster sampling and stratified sampling designs by helping to identify some of the important social and cultural characteristics of different groups which could not be obtained from the kinds of statistical sources normally used in sample selection. Exploratory research can also be useful in multistage sampling, where it is important to understand the composition of individual households within a multiunit building or compound.

Data Collection Methods

This issue has been discussed earlier. Data collection is the area in which the procedures and benefits of integrated approaches are best understood.

Understanding Context: The Political and Economic Environment, Project Implementation Processes, the Structure and Operation of Organizations, and Culture

The traditional approach to project planning and evaluation looked at *inputs, implementation processes* (how the inputs are used), *outputs,* and *impacts.* However, researchers are increasingly aware that projects take place in a certain context, and they are recognizing that it is important to understand the household characteristics, the socioeconomic environment, and the political and institutional environment within which the project is planned and implemented.

Although a quantitative survey is usually the best way to estimate the magnitude and distribution of poverty, the proportion of the population with access to different public services, or the quantitative impacts of projects, it is usually not the best method for understanding the socioeconomic environment, institutional and political processes, or how different kinds of households within different cultural contexts are going to respond to a project. Some researchers would go even further and challenge the conventional wisdom that survey research is an objective process for collecting "facts" about a community or activity. Instead they would argue that any kind of interview must be regarded as a social process in which the outcome is dependent upon the characteristics and expectations of the interviewer and respondents, and upon the context in which the interview takes place.

Qualitative methods, by comparison, are well suited for the analysis and interpretation of the context within which families live, or within which organizations or groups are operating and projects are implemented. Ethnography, sociology, and political science can evaluate contextual variables and explain how they affect the behavior and attitudes of the individuals or groups being studied. The analysis of these contextual and cultural factors should be an integral part of the research design. If this analysis is incorporated into an exploratory study during the research design phase, communities can be ranked on the contextual variables of interest to the study so that this information can be used in the design of stratified or cluster samples. However, qualitative methods of contextual analysis are often in-depth studies of a single or small number of communities or areas and, consequently, it may be difficult to generalize from these studies to assess the overall impact of these contextual factors at the regional or national level. In cases where it is necessary to generalize to large populations, the findings of the qualitative studies can be used to identify a few numerical indicators that can then be incorporated into quantitative surveys as part of large-scale studies.

Qualitative methods are particularly useful for the description of the project implementation process and for assessing the quality of implementation. Differences in outcomes and impacts for projects with similar resource endowments can often be attributed to differences in how the projects were implemented. Effective participatory approaches, adapted to local conditions, may be used in one project, while another project might use more rigid procedures that are less responsive to local conditions. As Sedlacek and Hunte (chapter 7) show, individual personalities (in this case the characteristics of the head teacher) can also have a major impact on project outcomes.

Qualitative methods can also be used to assess the quality of participatory planning and implementation methods. Every community and group has its own distribution of power, which is based partly on local traditions, partly on linkages to external political and economic systems, and partly on personalities. As a result, some participatory processes are dominated by a few traditional leaders or powerful people, while others are more open. It is frequently the case that certain groups, such as women, young people, economically weaker groups, or certain ethnic groups are at least partially excluded from effective participation in decision-making. Brown (chapter 8) reports that women were largely excluded from the planning and management of local water supply projects in Indonesia, even though water management is traditionally the responsibility of women. Surveys and methods relying on administrative information or the reports of community leaders are usually inadequate for capturing these political and cultural nuances.

Some of the recent World Bank poverty assessments (World Bank 1999) show how contextual and noncontextual methods can be combined with quantitative and qualitative data to explain how contextual factors such as local culture interact with standard indicators of poverty and welfare to influence responses to poverty alleviation programs as well as feelings of family members about their present and future conditions.

The following chapters provide examples of how qualitative methods have been used by researchers to clarify various contextual issues:

- *Chapter 4:* How the local political system affects allocation of resources in poverty alleviation programs.
- *Chapter 5:* What support mechanisms exist within different communities to protect families from fluctuating income and other household crises.
- *Chapters 6 and 7:* Understanding how participatory planning and decision-making takes place in the classroom under decentralized education, and how this affects the behavior of school directors, teachers, parents, and students.

- *Chapter 10:* How traditional male and female roles affect decision-making regarding the use of malaria control treatment.

Triangulation

Consistency checks and alternative measures of key variables should always be an integral part of integrated approaches. These alternative indicators should be used both as consistency checks and as a means of obtaining a deeper understanding of the variables being studied. Poverty research has shown, for example, that women and men may have a different understanding of the concepts of poverty. In some cases, Participatory Rural Appraisal (PRA) techniques such as wealth ranking produce rankings of the relative poverty of households or communities that are consistent with the survey estimates based on expenditure, consumption, or income, but in other cases the rankings may produce significant discrepancies. Women, for example, may place greater emphasis on the concept of vulnerability, which frequently concerns a lack of access to social support networks, than they do on current income or consumption (see the India poverty study in chapter 4). A widow, for example, may be ranked as more vulnerable to the consequences of drought or family crises than a married woman, even though the former may have a higher current income or expenditure.

The Indonesia water supply study (chapter 8) used triangulation to compare survey findings and observational estimates of the quality of community organization and the effectiveness of project implementation. Meetings were held to discuss and resolve any inconsistencies between the two estimates.

Data Analysis and Follow-up Field Work

The data analysis must incorporate both quantitative and qualitative analysis into an integrated analytical framework. The integrated sampling procedures discussed earlier must be used to assess the representativity of the findings from the case studies and the extent to which they can be generalized. The qualitative analysis of institutions, cultural variables, and the effectiveness of project implementation processes may be used to create dichotomous (dummy), nominal, or ordinal variables which can then be incorporated into the multivariate analysis to help analyze and explain differences in outcome variables in different communities or projects.

In the majority of research studies, inconsistencies between triangulated estimates are either ignored (the most common approach) or explained away in an unconvincing manner; and it is unusual to find a systematic treatment of inconsistent findings. A truly integrated analytical approach should include systematic procedures for identification and analysis of triangulation discrepancies and for addressing and resolving inconsistencies. The researcher should anticipate that such discrepancies will be found and should use them to enrich the analysis and interpretation. For example, systematic differences between the household income reported by women and men might reflect a weakness in the survey design, or it might provide some important insights into gender differences in control over, and information about, household resources.

Wherever possible the research design should include time and resources to permit a return to the field to follow up on these inconsistencies and to provide further insights into unexpected or difficult-to-explain findings. Rapid qualitative studies can make a valuable contribution in this area, because these methodologies are designed to provide greater depth of understanding concerning differences between communities, organizations, or households and to explain subtle differences that were not captured by the survey instruments. The study on village water supply in Indonesia (chapter 8) found that the water supply in one community was managed by men, whereas in all other cases this was the responsibility of women. The follow-up study revealed that women in this community were able to earn exceptionally high income from diary farming, and thus the men were willing to take over responsibility for water management in order to share in the earnings of their wives. Without the follow-up study, this unusual finding might have been ignored or even considered a reporting error.

Whenever possible, follow-up fieldwork should be conducted to

- investigate outliers (statistical findings which deviate from the normal pattern) to determine whether they represent a recording error or reflect an important new category. Examples include households or communities with higher or lower than expected incomes, deviations from the traditional sexual division of labor (see chapter 8 referred to in the previous paragraph), or unusual responses to economic crises;
- provide in-depth descriptions for each of the major categories or groups identified in the study (for example, households with different patterns of interhousehold network usage, discussed in chapter 5); and
- explain and illustrate the behavioral dynamics underlying the choice of economic or other analytical models.

Presentation of Findings

It is also important to ensure the integration of quantitative and qualitative data in the presentation of findings. The data gathered through case studies, wealth rankings, focus group reports, and other qualitative tools must be linked to, and compared with, survey findings. Case studies should be used not only to illustrate and enrich statistical tables but also to combine analysis on multiple levels. For example, how do cultural practices with respect to girls' education, water management, sexual division of labor, marriage, and other issues affect the observed statistical regularities concerning labor force participation, willingness to pay for community projects, the sustainability of infrastructure investments, and so on? The challenge is to capture the complex interactions between these different levels.

Table 1.2 proposes a set of guidelines for the design and implementation of a fully integrated, multidisciplinary research approach. A further discussion of how quantitative and qualitative methods can be used at each stage of the research process is provided in chapter 12.

Table 1.2 Elements of an Integrated, Multidisciplinary Research Approach

Research Team

- Include primary researchers from different disciplines. Allow time for researchers to develop an understanding and respect for each other's disciplines and work. Each should be familiar with the basic literature and current debates in the other field.
- Ensure similar linkages among national researchers.

Broadening the Conceptual Framework

- Draw on conceptual frameworks from at least two disciplines, with each being used to enrich and broaden the other.
- Ensure that hypotheses and research approaches draw equally on both disciplines. The research framework should formulate linkages between different levels of analysis.
- Ensure that concepts and methods are not taken out of context but draw on the intellectual debates and approaches within each discipline.
- Utilize behavioral models that combine economic and other quantitative modeling with in-depth understanding of the cultural context within which the study is being conducted.

Data Collection Methods and Triangulation

- Conduct exploratory analysis to assist in hypothesis development and definition of indicators.
- Select quantitative and qualitative methods designed to complement each other, and specify the complementarities and how they will be used in the fieldwork and analysis.
- Introduce contextual variables to help understand and explain attitudes and behavior.

(Table continues on the following page.)

Table 1.2 *(continued)*

- Select at least two independent estimating methods for key indicators and hypotheses.
- Utilize rigorous reporting standards for qualitative data collection and analysis procedures.

Sample Selection

- Define clearly whether, and how, qualitative findings are to be generalized.
- Where generalization to a larger population is required, ensure that appropriate statistical sampling methods are used for the selection of case studies.

Data Analysis, Follow-up Field Work, and Presentation of Findings

- Conduct and present separate analyses of quantitative and qualitative findings, and then show the linkages between the findings and levels.
- Utilize systematic triangulation procedures to check on inconsistencies or differing interpretations. Follow up on differences, where necessary, with a return to the field.
- Budget resources and time for follow-up visits to the field.
- Highlight different interpretations and findings from different methods and discuss how these enrich the interpretation of the study. Different outcomes should be considered a major strength of the integrated approach rather than an annoyance.
- Present cases and qualitative material to illustrate or test quantitative findings.

Chapter Overviews

Chapter 2. Issues and Approaches in the Use of Integrated Methods (Kimberly Chung)

This chapter presents a short introduction to research with qualitative methods and identifies some of the distinguishing characteristics of quantitative and qualitative methods as they are typically used at different stages of the research process. Chung stresses that there are very few clear distinctions between quantitative and qualitative approaches that hold in all cases. Whole volumes are written on qualitative methods; hence it is impossible to give a comprehensive overview in this chapter. The author addresses objections that are commonly cited as obstacles to accepting qualitative research and suggests reasons for using an integrated approach. Options for structuring an integrated research study are presented, along with prerequisites for conducting such research. The author cautions that no single approach will be best for all possible situations. The choice of methods will depend on the goals and budget of the study as well as the time and personnel available to it.

Chapter 3. Gender Issues in the Use of Integrated Approaches
(Roberta Spalter-Roth)

This chapter presents a sociologist's perspective on research methodologies, with an emphasis on approaches for studying gender issues. The author stresses the danger of creating a false divide between quantitative and qualitative methodologies and the need to avoid the assumption that only qualitative research can give women voice and provide insights on gender relations. She urges combining a variety of research strategies, including survey research done for other purposes, and provides examples of studies that use an integrated approach.

Chapter 4. Integrated Approaches to Poverty Assessment in India
(Valerie Kozel and Barbara Parker)

This chapter summarizes an innovative study of poverty in rural India that combined quantitative and qualitative research approaches. The authors discuss the objectives, methodological approaches, and preliminary findings of the study, with particular attention to issues such as underlying assumptions, sampling techniques, and lessons learned in the process of implementing an integrated research design. Recommendations for future research are also presented.

The study shows that poverty is an extremely complex phenomenon, and that an interdisciplinary approach is required to understand the sociocultural, political, economic, and institutional context within which people live their lives and which determines how poverty is experienced and perceived by different groups. Qualitative methods were essential for understanding the caste system and its role in determining attitudes to the possibility of escaping from poverty. The interdisciplinary approach also proved valuable in assessing the effectiveness of various government and donor-funded poverty alleviation programs, providing new insights into the strengths and weaknesses of some major initiatives.

Chapter 5. Studying Inter-Household Transfers and Survival Strategies of the Poor in Cartagena, Colombia
(Gwyn Wansbrough, Debra Jones, and Christina Kappaz)

This chapter presents the experience of using a combination of quantitative and qualitative research methods in a study of interhousehold transfers in the Southeastern Zone (SEZ) of Cartagena, Colombia, which was conducted as a follow-up to World Bank research on this topic conducted in the SEZ in 1982. The authors discuss the design and implementation of the study, which included both quantitative and qualitative approaches, and present their observations regarding integrated research. Issues addressed include survey design, sampling techniques, selection and training of interviewers, and timing and cost of research activities.

The researchers found that interhousehold transfers provided a significant proportion of the total household income, particularly for the poorest households and for female-headed households. The study demonstrated the value of an integrated approach in the analysis of interhousehold transfers. Transfers are determined by a complex set of social rules that are difficult to identify through formal surveys, and the transfers often take place in ways that are difficult to capture through surveys. The researchers concluded that once an exploratory qualitative study has been conducted to understand the dynamics of transfers in a particular community, it is possible to obtain reasonable estimates of the magnitude, distribution, and use of transfers through quantitative surveys. However, surveys will frequently fail to capture some of the transfers, and higher and more reliable estimates of the total volume of transfers will be obtained if surveys can be complemented by extended, informal interviews and participant observation with a subsample of households.

Chapter 6. Evaluating Nicaragua's School-Based Management Reform (Laura B. Rawlings)

This chapter discusses how a mixed-method approach was used to evaluate the impact of Nicaragua's school decentralization reform. Following an overview of the reform program and the objectives of the evaluation, the author discusses the quantitative and qualitative techniques utilized in the study, sampling issues, and the sequence of activities. The chapter summarizes the findings of the first phase of the study regarding the role of the school in governance, the perceived level of influence of key stakeholders (directors, teachers, and school council members), and the impact of the reform on school performance. This chapter contributes to the evaluation of the Nicaraguan reform by clarifying the objectives, methods, and value-added of the mixed-method approach.

Chapter 7. Evaluating the Impacts of Decentralization and Community Participation on Educational Quality and the Participation of Girls in Pakistan (Guilherme Sedlacek and Pamela Hunte)

This chapter focuses on the qualitative component of mixed-method research conducted in Pakistan to evaluate the impacts of decentralization and community participation on the quality of schools and to identify factors that influence the educational participation of girls. The authors discuss issues such as research design and sampling methods and present preliminary findings and their implications for the project. The authors also suggest a model for conducting follow-on studies in consultation with the community and offer recommendations for conducting integrated research.

The authors stress the importance of linking the selection of the areas for qualitative studies to the statistical sampling frame used for the quantitative studies in order to ensure the representativity and generalizability of the qualitative findings. The qualitative study was important in understanding what parents consider important in the assessment of a successful school. It suggested that the original project design might have placed too much emphasis on parental and community participation, and might have underestimated the importance of school organization, discipline, the critical role of the head teacher, and the importance of alliances between the head teacher and the community. While the integrated approach significantly increased the operational utility of the study, the costs and complexity of conducting this kind of research in remote areas such as northern Pakistan may limit the opportunities for replicating these research approaches in similar projects.

Chapter 8. Evaluating the Impact of Water Supply Projects in Indonesia (Gillian Brown)

This chapter describes integrated research methods that were utilized in the evaluation of rural water supply projects implemented by the Government of Indonesia with support from various multilateral and bilateral agencies. The author describes the research methodologies and data analysis techniques used in these studies and presents preliminary findings of Phase 1 research, including analysis of gender issues. Advantages and disadvantages of using an integrated research approach are discussed, along with recommendations for future research.

The samples for the qualitative work were selected to ensure their statistical representativity and comparability with the survey research work. Procedures were used to ensure that the rankings of communities (in terms of the effectiveness of project implementation and other variables) on the basis of survey reports were consistent with the more detailed qualitative assessments of the same communities.

The integrated approach proved valuable in several ways. First, it demonstrated a statistical correlation between the level and quality of community participation and the effectiveness and sustainability of implementation, and it provided an assessment of the quality of project implementation. Second, it questioned the conventional assumption that community contributions to the cost of a project are a good indicator of willingness to pay. In fact, many families were pressured by community leaders to make contributions even if they did not wish to do so. Third, it showed the need for qualitative measures of the level of women's participation, as very few women were actively involved in project management even though many indicated that they would have been interested in participating if they had been given the opportunity. Fourth, the qualitative methods provided an opportunity for rapid follow-up studies to assess unexpected outcomes, such as the fact that one of the best managed water projects happened to be the only one in which women were not involved in water management.

Chapter 9. Social Assessment of the Uzbekistan Water Supply, Sanitation, and Health Project (Ayse Kudat)

This chapter describes a social assessment process that was initiated in Uzbekistan as part of the preparation of a water supply, sanitation, and health project. The social assessments' focus on social development, participation, and institutional issues required empirical research using a combination of qualitative and quantitative methodologies. The author presents the objectives of this research and the range of methodologies that were used, including the decision to establish a local social science network. The findings of the impact of the various studies are also discussed.

Social assessment had a number of benefits as part of the overall planning and implementation process. First, it helped focus on the complementarities between the water supply, sanitation, and health sectors and helped develop integrated approaches covering all three sectors. Second, the social assessments helped the World Bank and many national agencies to understand the socioeconomic structure and its effect on how projects should be selected, designed, and implemented. Third, social assessments helped to understand the relationships between the Uzbeks and the Karakalpak regions and peoples, a major consideration in the design of projects that would affect both regions. Finally, the social assessments helped in conducting needs assessments for the more than 1,000 villages in the region and ensured that the proposed programs would address the priority needs of each subregion.

Chapter 10. Using Qualitative Methods to Strengthen Economic Analysis: Household Decision-Making on Malaria Prevention in Ethiopia (Julian Lampietti)

This study illustrates how qualitative methods can be used to enhance and complement the findings of quantitative survey methods. The chapter describes how qualitative methods are used at various stages of the research process to confirm the validity of quantitative results. The objective of the research was to measure the value people place on preventing malaria in themselves and members of their household in Tigray, a province in northern Ethiopia. While it is too early to discuss the actual valuation results, it is possible to highlight the interaction between qualitative and quantitative methods. An important methodological contribution was to show how qualitative methods can be used to provide the information and check the assumptions which will be built into the contingent valuation models. In particular, the qualitative studies were able to verify that the communities covered by the study fully understand the impacts of malaria on children and adults and were aware of the benefits and costs of currently available malaria prevention treatments.

Chapter 11. UNICEF's Use of Multiple Methodologies: an Operational Context (Mahesh Patel)

This chapter describes how UNICEF uses quantitative and qualitative methodologies to collect information for the purposes of monitoring, evaluating, and researching its programs and policies. Evaluation methodologies are selected according to the problem itself, the management decision faced, and the follow-through required. Complex problems may require use of multiple methodologies, some quantitative, some qualitative. A range of informational inputs and, hence, of evaluative methods may be needed for operational decisions and policy formulation. Required information can range from statistically valid survey data to opinions of leading politicians as obtained by key informant interviews. The most cost-effective evaluation strategy may be to optimally mix qualitative and quantitative methods, rather than use only one in a pure form.

Chapter 12. Lessons Learned (Michael Bamberger)

This chapter brings together all of the lessons learned from the workshop and the papers that were presented. It summarizes the benefits obtained from integrated approaches and shows how these can be implemented at each stage of the research process. The operational implications of integrated approaches with respect to cost, timing, and coordination are discussed, and some of the major challenges in using integrated approaches are identified.

The author is a senior sociologist in the Gender and Development Group of the World Bank.

Notes

1. I would like to thank my colleagues Nora Dudwick, Jesko Hentschel, and Lucia Fort who worked with me in organizing the workshop during which most of the material included in this report was presented. I would also like to thank Vijayendra Rao and Michael Woolcock for their helpful comments and for suggesting areas such as demographic and economic research in which the issues of integrated approaches have been extensively debated.

2. In this report referred to as "integrated research."

3. World Bank (1999).

4. Ibid, p. 47.

5. World Bank (1999), p. 48.

6. Hentschel (1999), pp. 88-89.

7. Pradhan and Ravallion (1998).

8. Ravallion and Lokshin (1999).

9. Mangahas (1995, 1999).

10. The poverty indicators used by the Social Weather Stations include asking families to rate themselves by selecting between cards saying "Poor" and "Not Poor," by indicating whether they have experienced hunger in recent months, and by indicating whether a member of the household has been sick during the past two weeks.

11. Riffault (1991).

12. See Cernea and Kudat (1997).

13. For example, *Population and Development* devoted the entire December 1997 edition to a symposium on the use of qualitative methods in what is normally considered to be a very quantitatively oriented discipline. On the other hand, Cluster Evaluation, which has been developed and popularized by the Kellogg Foundation, illustrates one approach to increasing the generalizability of participatory/empowerment evaluation (Sanders 1997).

14. Miles and Huberman (1984).

15. Bamberger and Kaufmann (1984).

16. Obermeyer (1997).

17. "The increasing popularity of qualitative methods as a complement to traditional demographic analyses, and the growing taste among demographers for such tools as open-ended interviews, focus groups, and rapid assessment, have evoked among knowledgeable observers a mixture of enthusiasm and trepidation . . The trepidation because the uncritical use of qualitative methods may lead to superficial analysis and simplistic views." (Obermeyer 1997)

18. Kertzer and Fricke (1997).

19. The definition of QL as nonquantitative is important, because demography is considered a rigorous mathematical field. Many demographers limit analytical frameworks to what can be quantified so that there are "constraints on the demographic imagination." (Obermeyer 1997)

20. For example, Rao (1997) and Brewer and Hunter (1989).

21. Kertzer and Fricke (1997); Obermeyer (1997).

22. Ragin (1994: pp. 107-8) (cited in a personal communication from Michael Woolcock).

23. Datta (1997).

24. Obermeyer (1997, p. 815).

25. Obermeyer (1997).

26. Kertzer and Fricke (1997).

27. Rao (1997).

References

Bamberger, Michael, and Daniel Kaufmann. 1984. "Patterns of Income Formation and Expenditures among the Urban Poor of Cartagena." Final Report on World Bank Research Project No. 672-57, The World Bank, Washington, D.C.

Bamberger, Michael, Mark Blackden, and Abeba Taddese. 1994. *Gender Issues in Participation.* Environment and Socially Sustainable Development Department. The World Bank.

Bamberger, Michael, and Jerry Lebo. 1999. Gender and Transport: A Rationale for Action. PREMNOTE No. 14 Poverty Reduction and Economic Management Network. The World Bank.

Barwell, Ian. 1996. *Transport and the Village.* World Bank Discussion Paper No. 344. The World Bank.

Blackden, Mark, and Elizabeth Morris-Hughes. 1993. *Paradigm Postponed: Gender and Economic Adjustment in Sub-Saharan Africa.* AFTHR Technical Note No. 13. Africa Human Resources Department.

Brewer, John, and Albert Hunter. 1989. "Multimethod Research: A Synthesis of Styles." Sage Library of Social Research 175. Thousand Oaks, Calif: Sage Publications.

Bryceson, Deborah Fahy, and John Howe. 1993. *African Rural Households and Transport: Reducing the Burden on Women?* IHE Working Paper IP-2. International Institute for Hydraulic and Environmental Engineering. Delft: The Netherlands.

Cernea, Michael, and Ayse Kudat, eds. 1997. "Social Assessments for Better Development: Case Studies in Russia and Central Asia." Environmentally Sustainable Development Studies and Monographs Series 16. Washington, D.C.: The World Bank.

Datta, Lois-ellin. 1997. "Multimethod Evaluations: Using Case Studies Together with other Methods." In *Evaluation for the 21st Century,* ed. Eleanor Chelimsky and William Shadish. Thousand Oaks, Calif.: Sage Publications

Hentschel, Jesko. 1999. "Contextuality and Data Collection Methods: A Framework and Application to Health Service Utilization." The Journal of Development Studies. Vol. 35, No. 4, April 1999, pp. 64–94.

Hentschel, Jesko. 1998. *Distinguishing Between Types of Data and Methods of Collecting Them.* Policy Research Working Paper 1914. Washington, D.C.: The World Bank.

Kertzer, David, and Tom Fricke. 1997. "Toward an Anthropological Demography." In *Anthropological Demography: Towards a New Synthesis,* ed. D. Kertzer and T. Fricke. University of Chicago Press.

Mangahas, Mahar. 1999. "Monitoring Philippine Poverty by Operational Social Indicators." Paper presented at the World Bank's Poverty Reduction and Economic Management Week, July 14, 1999.

Mangahas, Mahar. 1995. "Self-Rated Poverty in the Philippines: 1981–1992." *International Journal of Public Opinion Research* 7:1.

Miles, Matthew, and Michael Huberman. 1984. *Qualitative Data Analysis: A Sourcebook of New Methods.* Thousand Oaks, Calif.: Sage Publications.

Obermeyer, Carla Makhlouf. 1997 "Qualitative Methods: A Key to a Better Understanding of Demographic Behavior?" In Qualitative Methods in Population: A Symposium, ed. Carla Makhlouf Obermeyer, *Population and Development Review* 23(4): 813–18.

Pradhan, Menno, and Martin Ravallion. 1998. "Measuring Poverty Using Qualitative Perceptions of Welfare." Policy Research Working Paper No. 2011. Development Research Group, Poverty and Human Resources. Washington, D.C.: The World Bank.

Ragin, C. 1994. *Constructing Social Research: The Unity and Diversity of Method.* Thousand Oaks, Calif., and London, England: Pine Forge Press.

Rao, Vijayendra. 1997. "Can Economics Mediate the Relationship Between Anthropology and Demography?" In Qualitative Methods in Population: A Symposium, ed. Carla Makhlouf Obermeyer, *Population and Development Review* 23(4): 813–18.

Ravallion, Martin, and Michael Lokshin. 1999. "Subjective Economic Welfare." Policy Research Working Paper No. 2106. Development Research Group, Poverty and Human Resources. Washington, D.C: The World Bank.

Riffault, Helene. 1991. "How Poverty Is Perceived." In *Eurobarometer: The Dynamics of Public Opinion,* ed. Karlheinz Reif and Ronald Inglehart. London: Macmillan Academic and Professional Limited.

Sanders, James. 1997. "Cluster Evaluation." In *Evaluation for the 21st Century*, ed. Eleanor Chelimsky and William Shadish. Thousand Oaks, Calif.: Sage Publications.

World Bank. 1999. *Poverty Reduction and the World Bank: Progress in Fiscal 1998.* Washington, D.C.: The World Bank.

2
Issues and Approaches in the Use of Integrated Methods

Kimberly Chung

This chapter presents a short introduction to the use of qualitative methods for collecting data. It draws comparisons between the collection of structured survey data—a process with which many workshop participants were familiar—and less-structured, qualitative approaches to data collection. The author addresses some common objections that positivist survey researchers (particularly economists) raise about qualitative data or qualitative analyses. Options for structuring an integrated study are presented, along with prerequisites for conducting such research. The author cautions that researchers experienced with collecting and analyzing qualitative data hail from diverse epistemological backgrounds. As such, it is common to encounter differences of opinion on the purpose, uses, and analysis of qualitative data. In addition, as with all other methods of data collection and analysis, no single approach will be best for all possible situations. Ultimately, the choice of method will depend on the goals of the study as well as the budget, time, and personnel available to it.

To my knowledge, there is no clear consensus on the definition of "qualitative data." Different researchers assume different definitions, and most are not explicit about what they mean when they use the term. Nor does this chapter offer a perfect definition. However, as a researcher trained in the traditions of experimental science and survey research, I have often been asked to explain qualitative approaches to people whose backgrounds are similar to mine. I often respond by discussing the ways in which qualitative data differ from data collected through controlled experimental studies or highly structured survey studies, such as the Living Standards Measurement Studies (LSMS).

Volumes have been written on qualitative data collection methods, and it is impossible to give a comprehensive overview in such a short presentation. I also recognize that many researchers will disagree with the definition that I give below. However, I think that this description is a useful starting place for those accustomed to the form of survey research that is commonly conducted at the World Bank. It is with these caveats that I approach the questions of "What are qualitative data?" and "How can we integrate their collection and use with the process of collecting and using survey data?"

What Are Qualitative Data?

In my view, qualitative data are defined by the *format* of the data and the *process* by which they are generated.[1] In terms of format, qualitative data are often textual records. These records draw heavily on context, local perceptions, and a holistic understanding of the phenomenon under study. In their raw form, qualitative data often consist of notes prepared by the researcher or tape recordings of interviews or conversations with subjects. For example, a researcher's notes may include physical descriptions, observations of community activities (such as funerals, community work projects, or political events), or responses from an interview. Most qualitative data are collected *de novo* for a given study, but secondary data such as existing written reports (for example, minutes of a meeting) may also be used. Less frequently, other media such as photographs may be used.

Qualitative data are also defined by the process through which they are generated. To my mind, methods for obtaining qualitative data are less structured than those used in large sample surveys (such as the LSMS). These methods include questions that permit greater flexibility in the manner in which respondents provide answers. For example, a typical sample survey asks respondents to provide answers in a format that is rigidly controlled by the investigator (e.g., a short, codable response or numerical response). By comparison, qualitative methods tend to use more open-ended questions that get at the "why" and "how" underlying a phenomenon. For example, a researcher might want to understand the diverse incentives that a person considers when undertaking a certain action (e.g., taking a high interest loan). Since it may be difficult for respondents to explain these incentives fully with only a set of precoded answers, they are permitted to give longer answers using their own words. The unstructured response also allows for the possibility of discovering new explanations or relationships that researchers had not known to be important a priori. Since the questions are more open-ended, interviewers will usually probe further to gain clarification or to focus the respondent on specific elements of the answer that are more interesting to the study. Usually this entire exchange between the researcher and the respondent would be recorded in textual form as data.

This process is quite different from that typically used in survey research. Surveys are highly structured and usually use closed-ended questions that focus on "what?" and "how much?" In addition, when discussions occur between an investigator and a respondent, little information other than the final numerical answer or code is recorded on the survey. As a result, survey data are almost always collected in a numerical or coded format.[2] The principal advantage of such surveys is that they can be administered to large numbers of individuals, organizations, or households using standardized methods. Standardization across observations makes it possible to aggregate measures and to make statistical comparisons among individuals, households, regions, and time periods.

In addition to the format of questions and answers, qualitative data collection differs from that used for a survey in that researchers are usually less concerned about rigidly applying their data collection protocol to each unit of observation. With a structured survey, it is important to administer each questionnaire identically; that is, exactly according to a predetermined protocol. However, the data collection protocol for qualitative studies can be relatively flexible in comparison to that for a survey because there is not always a need to directly compare the answers given by each respondent. In addition, qualitative researchers typically start to analyze and digest new data as they are collected in the field. As a result, they can capitalize on what is learned in the field and make immediate adjustments to the data collection protocol. Therefore, a qualitative researcher might modify the mode of data collection while it is in progress in order to pursue certain preliminary findings.

This is not to say that qualitative data are always unstructured or that they are "undisciplined." To the contrary, qualitative protocols can vary in their degree of structure. For example, in cases where results are to be compared across sites, interviewers might be asked to follow a precisely defined protocol so that the data remain comparable. As another example, certain genres of qualitative research, (e.g., grounded theory) have fairly structured expectations of how data collection and analysis should proceed. The point is that qualitative data collection is often, but not always, more flexible and less rigidly structured than survey data collection.

There are other differences. Qualitative data collection often uses sampling techniques that are different than those commonly employed for the collection of survey data. Typically, survey research relies on some form of random sampling to ensure that the results can be generalized from the sample to the population. Qualitative protocols use random sampling when it is appropriate, but they also make use of nonrandom methods of sampling. For example, it may be advantageous to focus on unusual cases rather than typical cases. A researcher conducting a study on the care-giving determinants of poor preschooler nutritional status might note that children living in a high-income household, who one would expect to be well nourished, were unusually thin and stunted. This finding might lead the researcher to conduct a case study of that household to try to uncover the factors responsible for this unexpected outcome. On the other hand, case studies may also be prepared on typical households or communities to help characterize the mean tendencies.

Qualitative Data Collection Methods

There are many different types of qualitative data collection methods. These include:

- *Informal or semistructured interviews with individuals,* using open-ended questions.
- *Focus groups.* Ideally, these are comprised of small, homogeneous groups formed by the researcher to discuss open-ended questions about a certain topic. Focus group respondents are encouraged to talk among themselves so that a discussion unfolds among the participants rather than between the researchers and the respondents.
- *Community meetings,* which are somewhat larger and less homogeneous than focus groups but follow a similar format. These meetings take on a more "question-answer" format than a focus group, with the researcher clearly asking questions and members of the group responding.
- *Direct observation of events, behaviors, or physical structures.* Direct observation consists of the investigator's own observations and perceptions, as opposed to those of a respondent.
- *Systematic data collection methods, such as pile sorts, triadic comparisons, and ranking.* These methods ask respondents to make comparisons among various items. The comparisons might be made on paper (as with a self-administered survey), or by means of grouping cards according to the pictorial cues found on them. In either case, the data are collected in a way that permits numerical or qualitative analyses. These methods do not strictly meet the definition of qualitative data given above. I mention them because they may be used with open-ended qualitative techniques and they are not used widely at the World Bank.
- *Methods Popularized by the Participatory Research Movement.* Methods used in participatory research are designed to promote the active participation of the individuals or communities in the research process. The purpose is often to give voice to politically or socially vulnerable groups who do not normally have the opportunity to express their views. Often the research is conducted with groups and the researcher seeks to play the role of facilitator rather than that of data collector. Nonverbal, pictorial techniques (such as constructing maps, histograms, or ranking charts) are commonly associated with participato-

ry research. With or without such techniques, participatory research relies on promoting discussion in which group members participate actively.

A Comment on Participatory Research

A common mistake among those unfamiliar with participatory research is to equate the use of the pictorial methods (which are often used in participatory research) with the act of conducting participatory research. It is important to underscore that the methods popularized by the participatory literature might be used for very different goals, some which are more participatory than others. As an example, such methods might be used as part of an action program that raises the awareness of vulnerable groups (e.g., women, the elderly, or indigenous groups) about their political and social environment. The long-term goals of this program might be to empower community members to identify issues that are important to them and to plan for change (or action) on these issues. In this case, the participatory element of this study is the active engagement of the community in the process of change. In an ideal, highly participatory situation, community members would be involved in setting the terms of the research—deciding what questions they want to explore, what information they need to collect, and how the information would be used.

By contrast, these same methods of data collection may be used in a completely different spirit. That is, to collect data that reflect the opinions and experience of the people studied, but for purposes that serve the interest of the researcher and not the participants. In such a situation the terms are set completely by the researcher. The researcher decides on the type of information that is needed and how it will be used. In this situation, if the study is not tied to an action plan or to any ongoing work with the community, it is not likely to lead to any sustained sense of empowerment or change within the community. This is because the respondents are involved in the process only to the extent that they are providing answers to a researcher's questions. In terms of the quality of the participation, there is little to distinguish this form of data collection from others.

I believe that the extent to which respondents' voices are able to affect policy or programming decisions determines whether such a study is participatory or not. If a method (qualitative or otherwise) is used to collect information that serves only the needs of the researcher, there is little that is participatory about the process. On the other hand, if the research incorporates the views of the community (or target group) in a process that leads to changes they value, then I believe the process has been participatory—*regardless of the actual data collection method used.* In short, any method can be used in a participatory manner; it's the spirit and the purpose for which it is used that determines the extent to which the research is participatory.

This suggests that there are varying degrees and qualities of participatory research, and that there is no single definition of what is truly "participatory." One workshop participant noted that including feedback sessions to the community that report on the results of a study can be appropriate and empowering and thus participatory, even if it is not directly linked to change within the community. She stated that this is particularly true in situations where there has been no tradition of democracy or where there is a strong hierarchy within the community that traditionally limits access to or input from certain groups. I agree but maintain that it is important to assess whether the subjects are gaining anything by participating. Repeated studies, even those that use participatory methods, may cease to be empowering if they don't lead anywhere—if participants do not feel that they have benefited from the process. Of course, much of this will depend on the nature of the relationship between the researcher and the participants/communities.

Participation is difficult to achieve and requires patience and sensitivity to the local context. One workshop participant noted that it is particularly difficult within communities with strong dynamics of social exclusion that are incompatible with the goal of creating equity.

Some Problems That Survey-Based Researchers Cite as Barriers to Accepting Qualitative Work

Researchers who are experienced only with surveys are frequently uncomfortable with qualitative data and the methods used to analyze qualitative data. Some of the commonly cited obstacles to accepting a qualitative approach are listed below:

- Qualitative research is subjective—how do I know if the analyses or the data are credible?
- How can qualitative data be used for anything besides a "box story" to accompany the quantitative analysis?
- Qualitative research costs too much.
- There's nothing I can learn from asking local people about this topic.
- The data can't be generalized.

Several of the following chapters argue that one way for survey researchers to increase their understanding of, and confidence in, qualitative methods is to hire people they trust and respect intellectually and who share the same goals for the study. Many qualitative methods are foreign to conventional survey researchers, and it is important to include in the research team qualitative researchers who sufficiently understand conventional survey methods in order to provide a link between the two approaches. It is also important to use a team approach in developing the goals of the qualitative research and to collaborate closely during all phases of the research.

This does not, of course, overcome concerns about subjectivity and rigor, some of which are based on epistemological differences among researchers. Yet many of these concerns can be overcome by making the methods more transparent so that the logic behind data collection and data analysis decisions is apparent. One way to do this is to require sufficient documentation of the methods by which data were collected and analyzed.

The issue of how to ensure that qualitative data can be used for more than a "box story" is best addressed by suggesting ways in which qualitative research can enhance survey research or program development. The following section identifies a number of benefits from the use of integrated approaches, including improved quality of data collected through surveys and improved responsiveness of programs to the needs of the target population. The key is to have clear goals for the qualitative component and to integrate it closely with the quantitative work from the outset. The goal might be to answer a very specific question or to do some exploratory work in an area in which very little is known. In either case, it is important to know at the outset why you wish to collect qualitative data and to organize the study accordingly.

Reasons to Integrate Qualitative and Quantitative Research Approaches

There are at least two major reasons for integrating qualitative research with survey research at the World Bank. The first is to use qualitative work to improve the quality of data collected through surveys. The second is to use qualitative work to improve work in operations, that is, to improve the responsiveness of programs to the needs of target groups by making them more participatory.

Using Qualitative Data to Improve Research Conducted with Surveys

I would like to highlight three different ways in which qualitative data can improve survey research. The first is to use qualitative work to identify hypotheses that might be tested with survey data. For example, researchers may want to study intrahousehold resource allocation in a particular area, but they may not know the relevant issues. Preliminary qualitative work using focus groups, unstructured interviews, observation, or other techniques would help the researchers identify the key issues. Although most survey researchers are familiar with this approach, this is the first place where cuts are made to a study when time or money is limited. In addition, exploratory work is often not done very systematically because many survey researchers are not aware of the range of qualitative techniques that are available. However, the literature contains many examples of researchers who enriched the quality of their surveys because the qualitative research enabled them to include variables or propose hypotheses that they would otherwise have overlooked.

Another area in which qualitative work is useful is to improve the quality of the survey instrument—in other words, to improve the clarity of terms and codes and to come up with appropriate questions to ask. Survey questionnaires are often based on instruments that have been used successfully in other locations but have not been tested locally. In many cases, the questions in the modified instrument are highly confusing to respondents. Qualitative research can help the researcher understand how people think about the concepts that are being studied, how they would categorize things, how they talk about the issues, and how the survey questions should be phrased.

Third, after a survey has been conducted, qualitative research can help to confirm hypotheses. For example, a researcher may be unable to explain counterintuitive or inconclusive findings of the quantitative analysis because the correct variables were not included in the survey or they were not sufficiently disaggregated to allow for the appropriate analysis. Qualitative work can be useful in this case, because the researcher can go back to the field and ask focused questions to validate various hypotheses or to seek explanations for counterintuitive findings. It can also be used to sample from the extremes to learn more about cases that differ significantly from the norm.

Using Qualitative Methods to Improve Responsiveness of Programs

In addition to improving survey research, qualitative methods can be used to improve the responsiveness of programs to clients. There are two ways to think of this. First, qualitative research can be used to understand the needs of the client groups or to understand the nature of their situation. This requires a shift in philosophy and a recognition that the client group may know what is good for them. In the broader picture, this means accepting that there is no single interpretation of "truth." For example, the researcher's understanding of what it means to be poor may be different from how the poor think about it and experience it. Again, quantitative and survey methods can be used either sequentially or simultaneously to accomplish this objective. Qualitative research can be conducted as a first step to learn more about poverty and how people conceptualize it, and the data can then be used to develop a survey that can quantify and locate the poor according to these criteria. One caveat is that location-specific definitions may not aggregate. This will be problematic if the research includes a very large sample or if the study is conducted on a national level.

Another approach that has been used very successfully by nongovernmental organizations is to use qualitative or participatory research to design programs for a particular target group. For example, surveys can be used to identify groups that are income poor, and then a qualitative or participatory method can be used to design or evaluate the effectiveness of a program. A potential problem in this case, however, is that the researcher is starting with the outsider's definition of the target group.

Integrating Qualitative and Quantitative Methods

Qualitative and survey methods can be integrated either sequentially, in which one component of the research is conducted after the other is completed, or simultaneously, in which the two are conducted in parallel. I believe the sequential approach is easier to manage and can be more cost efficient. Ultimately, it all depends on the goals and scale of the study. It may make sense to use a simultaneous approach in situations in which the two types of research are used to inform each other. However, this requires significant coordination if the two approaches are to be integrated, and it can be costly (in terms of researcher time), since researchers must be in the field to make these adjustments.

Prerequisites for Conducting Integrated Research

The first prerequisite for survey researchers conducting integrated research is to have an open mind. Researchers must be willing to draw on a wider range of data collection and analytical methods, to learn from collaborators who bring different perspectives to the work, and to learn from the people interviewed in the field.

Second, the researcher needs a collaborator who can do the qualitative work, one whom he/she respects intellectually, and who shares an interest in working with people with different perspectives.

Third, the research team needs to agree upon a set of goals for the qualitative component of the research. This helps to focus and coordinate the work and to keep the costs down.

Fourth, there must be adequate financial resources to involve the qualitative researcher in all stages of the research, from setting the goals to collecting and analyzing the data. The best results happen when teams work together from the outset.

Finally, the quantitative researcher must have a willingness to work with the qualitative researcher and to integrate the qualitative findings with the survey data as a core part of the research.

The author is an assistant professor of community and economic development at Michigan State University.

Notes

1. Survey researchers may consider bivariate or categorical data to be qualitative. Since most workshop participants were familiar with how these data are collected and analyzed, this discussion is focused on an alternate definition of "qualitative."

2. It is, of course, possible to include open-ended questions, which can be analyzed qualitatively, in a survey. In my experience, however, this is uncommon in surveys written by economists or scientists.

3
Gender Issues in the Use of Integrated Approaches
Roberta Spalter-Roth

This chapter presents a sociologist's perspective on research methodologies, with an emphasis on approaches for studying gender issues. The author stresses the danger of creating a false divide between quantitative and qualitative methodologies. She emphasizes the importance of employing methodologies that not only give underrepresented women a voice in the policy process but provide valid measures of persistent gender inequalities. She urges combining a variety of research strategies, including survey research done for other purposes, and provides examples of studies that use an integrated approach.[1]

Sociologists and Research Methodologies

The American Sociological Association has 13,000 members who do both quantitative and qualitative research. The 1997 edition of one of the association's annual publications, *Sociological Methodology*, had a major section on improving the quality and rigor of qualitative research.[2] Sociologists have been debating quantitative versus qualitative approaches forever. Students of sociology learn both methods, and each is considered part of the research craft. Consequently, although you can always strike up a lively debate on this topic, for many sociologists the legitimacy of using both quantitative and qualitative data to increase the validity of studies, and the interdependence of these research strategies for producing valid knowledge, is by and large no longer an issue. The debate is periodically revisited during the formation of new theoretical areas of the discipline, new problems, and new paradigms. Most recently, the growth of gender analysis as a major research area within the discipline resulted in a resurgence of this debate, as researchers attempt to determine the best methodologies for both understanding and overcoming relations of gender inequality.

Currently, the section on Sex and Gender is the largest interest group within the American Sociological Association. Among the many researchers working in this area, there is a consensus that researchers should use the best method available—whether quantitative or qualitative—to understand gender as a basic organizing principle of social life. Moreover, it is seen as essential to examine both *structure* and *agency*. An analysis of both focuses on how people operate within structural context, and how they use those structures in ways that may not have been anticipated by those who plan and most benefit from them.

Avoiding False Divides

The question of whether quantitative research is preferable to qualitative research creates a false divide for researchers. It is particularly important to avoid subscribing to the notion that quantitative research is "scientific," "hard," "masculine," and "objective," while qualitative research is "interpretive," "soft," "feminine," and "subjective." This perception leads to the unfortunate conclusion that research done by women is by definition qualitative, and that only qualitative research can understand and give voice to women. Qualitative research should not be equated with gender analysis, and quantitative research should not be equated with universalistic analysis. All research, regardless of method or technique, must deal with constant tradeoffs between cost, time, validity, generalizability, completeness, and resonance with both those who are the subjects of the research and those who will read the research. Issues such as power relations between researchers and subjects, empowerment of subjects, links between studies and practice, and involvement and detachment of researchers in practice are critical for all researchers, regardless of the methods used. It is important that all researchers use two types of indicators when they investigate broad social issues such as poverty; race, ethnic, and gender discrimination; and social and economic restructuring These are, first, indicators of the standpoints and actions of the actors in the situation, especially the often left-out voice of poor women (agency), and, second, indicators of change or lack of change in relations of persistent inequalities such as those between employers and workers, landlords and tenants, or husbands and wives (structure).

My own experience in conducting research to bring about social change and influence social policy indicates that the most persuasive policy research includes both of these elements: *numbers* that define the scope and patterns of the problem, and a *story* that shows how the problem works in daily life and provides for empathetic understanding. These two elements stem from quantitative and qualitative research. Integrated approaches are as effective for the analysis of victims' stories of exploitation and oppression as they are for the analysis of "success stories" of programs, policies, or actions that helped to overcome persistent inequalities.

Improving Quantitative Data for Studying Gender Issues

In spite of the tendency to equate gender analysis with qualitative methods, a significant amount of research on gender issues in the United States has been quantitative. This research has used major sample surveys such as the Panel Study of Income Dynamics, the Survey of Income and Program Participation, the National Longitudinal Survey, the Current Population Survey, and the Integrated Public Use Microdata Series. These surveys were not originally designed to examine women's status or gender relations. Nonetheless, they can be used for that purpose, depending upon the questions that are asked and the focus of the research. Researchers should be working to improve the validity of quantitative surveys so that they better reflect the patterns and processes of women's lives. Because many feminist researchers raised these issues, quantitative instruments such as the Current Population Survey, for example, now do a better job of measuring women's labor force participation. Fewer U.S. data sets, for example, automatically assign headship to the husband and father in the household. These changes can serve as models for further refinements.

Some of the emphasis on qualitative research, especially in the areas of women's informal work and inter- and intrafamily relations, results from the unwillingness to fund large quantitative studies that systematically measure these relations. For example, when development organizations began to study gender issues, there was often an assumption that this analysis could be studied through small-scale qualitative research or case studies, because activities such as housework are too difficult to measure.

In fact, there are hundreds of studies that illustrate appropriate quantitative methods even under circumstances in which women themselves do not consider their housework as work.[3] A critical set of pioneering studies to devise methods for measuring persistent inequalities in work and consumption, when the victims themselves perceived these inequalities as part of the nature of things, are found in *Tyranny of the Household: Investigative Essays on Women's Work*, edited by Devaki Jain and Nirmala Banerjee.[4] These studies used population censuses, participant observation, questionnaires, time diaries, and measures of food consumption to quantify inequalities between adult men and women and boys and girls.

In order to go beyond simply including gender as an unspecified independent variable, researchers must focus on questions of how persistent gender inequalities are produced and reproduced. Women and gender relations must be the central focus of the research. It is the research question, not the method, that is important. Although obvious, it should be noted that persistent gender inequalities are not limited to third world countries. In some of my own research on low-income women and their families, I used nationally representative data from the Survey of Income and Program Participation—which includes information on family composition, labor force participation, and income sources—to develop active portraits of women who "packaged" income from a variety of sources, including state welfare payments, to bring their families out of poverty. This research focused on the activities of women rather than viewing them as passive recipients of welfare.[5]

Giving Voice to Women

Feminist epistemology maintains that the researcher is not an objective neutral person but a contextual, historical being whose views can affect the research. It also holds that the subjects of the research can ask questions of researchers, which often leads to the development of more valid research. A variety of path-breaking books include extensive discussions of methods, the role of the researcher and the researched, and the issue of giving voice.[6] Although it is important to give women voice in research studies, and efforts to do so must be encouraged, it is equally important that researchers develop other measures of persistent inequalities and how they are produced and reproduced. Giving voice to women's understanding of their lives through qualitative research may or may not help to analyze persistent inequalities, because women's lack of power may not automatically produce individual understandings of institutionalized patterns, and inequalities which are often invisible. It is important to understand that in order to overcome these inequalities, researchers must go beyond "giving voice" in studies.

In recent years there has been a tremendous growth in independent women's policy research organizations. Researchers at these organizations are trained to do research that puts women and their daily experiences at the center. In Barbados, for example, there is an excellent Women in Development program that conducted extensive quantitative and qualitative research during the Decade on Women, when money was available for such work.[7] These groups should be funded to do research that gives voice to women and their daily experiences, analyzes persistent inequalities, and relates research to practice.

Integrated Approaches to Gender Research

Two recent policy research studies done in the United States show how integrated approaches can provide indicators of women's standpoints and actions under conditions they did not create. They are examples of studies that give women voice and use integrated methods to provide valid and reliable findings.

The first is a 1997 study by Kathryn Edin and Laura Lein on how welfare mothers survive on below-poverty welfare payments.[8] The researchers began by using the Consumer Expenditure Survey, through which they found that expenditures in some households were significantly greater than reported income. To learn more about the dynamics behind that finding, they used a snowball sample, getting referrals through trust networks, and began to spend time with these women. To make the subjects more at ease and to create the ambiance of a mother-to-mother conversation, one researcher brought her own children along to the interviews. This qualitative work showed that some women had other sources of income, such as boyfriends, or were more able to do off-the-books work than others. These women and their families could not survive without this additional income, and they suffered many hardships despite it.

This is not the type of information that people give their local census enumerator; it could only be obtained through in-depth interviews after some relation of trust was established. The original research was a small qualitative study, but the results were so compelling that the Russell Sage Foundation sponsored a four-city replication to increase the validity and generalizability of the findings. The study was so powerful that one of the researchers was featured in a cover story in the *New York Times Sunday Magazine* telling a sympathetic story of these women's lives.

The second example is a 1994 study called *Working Women Count* that was conducted by the Women's Bureau of the U.S. Department of Labor. The purpose of this study was to learn how women felt about their work lives and what they saw as the major issues and problems. Under the direction of Karen Nussbaum, then director of the Women's Bureau, the researchers used two different data-gathering techniques. First, they developed a series of partners—in total more than 1,600, including businesses, grass-roots organizations, unions, daily newspapers, national magazines, and federal agencies in all 50 states, the Virgin Islands, Guam, and Puerto Rico—and asked these organizations to distribute questionnaires to their workers. They received nearly a quarter of a million completed questionnaires in return. The researchers then conducted a telephone survey with a scientifically selected, national random sample to ask the same set of questions.

Despite the use of different data-gathering techniques, there were very few differences between the two groups' responses. The majority of respondents wanted more respect, better pay on their jobs, higher valuation of the work they did in caring for their families, and better government and corporate work-family policies. This research was subsequently used to develop a governmental policy agenda based on the respondents' views of their lives, their issues, and their problems.

Conclusion

Quantitative and qualitative research should not be regarded as polar opposites. Both can and should be used to give women voice, to analyze persistent, institutionalized gender inequalities, and to encourage the overcoming of inequalities by placing these issues on the policy agenda. The decision of which research method to use depends on the research questions and the time and resources available. The association of qualitative research with gender issues should not be used as an excuse not to fund large-scale studies that put women and their daily lives at the center of the research. Because of their experience in dealing with these issues, sociologists should be called upon to assist with efforts to integrate research approaches.

Roberta Spalter-Roth is director of research at the American Sociological Association.

Notes

1. This chapter is based on a brief talk delivered as part of a panel. It was not meant to do justice to the topic of gender issues in the use of integrated approaches. It does not represent the view of the American Sociological Association.

2. Rafferty (1997).

3. See, for example, Goldschmit-Clermont (1987).

4. Jain and Banerjee (1985).

5. Spalter-Roth, Burr, Shaw, and Hartmann (1995).

6. See, for example, Gottfried (1996) and Reinharz (1992).

7. See, for example, Mohammed and Shepherd (1988).

8. Edin and Lein (1997).

References

Edin, Kathryn, and Laura Lein. 1997. *Making Ends Meet: How Single Mothers Survive Welfare and Low-Wage Work.* New York: Russell Sage.

Goldschmit-Clermont, Luisella. 1987. *Economic Evaluations of Unpaid Household Work: Africa, Asia, Latin America, and Oceania.* Geneva: International Labor Organization.

Gottfried, Heidi, ed. 1996. *Feminism and Social Change: Bridging Theory and Practice.* Urbana: University of Illinois Press.

Jain, Devaki, and Nirmala Banerjee, eds. 1985. *Tyranny of the Household: Investigative Essays on Women's Work.* New Delhi: Shakti Books.

Mohammed, Patricia, and Catherine Shepherd, eds. 1988. *Gender in Caribbean Development.* Cave Hill, Barbados: University of the West Indies, Women in Development Project.

Rafferty, Adrian E., ed. 1997. *Sociological Methodology* 27. Oxford: Basil Blackwell.

Reinharz, Shulamit. 1992. *Feminist Methods in Social Research.* New York: Oxford University Press.

Spalter-Roth, Roberta, Beverly Burr, Lois Shaw, and Heidi Hartmann. 1995. *Welfare that Works: The Working Lives of AFDC Recipients.* Washington, D.C.: Institute for Women's Policy Research.

U.S. Department of Labor Women's Bureau. 1994. *Working Women Count.* Washington, D.C.: Department of Labor.

Part II
Lessons from the Field

Poverty Analysis

4
Integrated Approaches to Poverty Assessment in India[1]

Valerie Kozel[2] and Barbara Parker

This chapter summarizes an innovative study of poverty in rural India that combined quantitative and qualitative research approaches. The authors discuss the objectives, methodological approaches, and preliminary findings of the study, with particular attention to issues such as underlying assumptions, sampling techniques, and lessons learned in the process of implementing an integrated research design. Recommendations for future research are also presented.

Research Objectives

An innovative study of poverty in some of India's poorest regions has recently been implemented. The primary objective of this study, which combines quantitative and qualitative research approaches, was to learn more about poverty in rural India in order to provide input into the next phase of poverty work for the India country program of the World Bank.

Although much is already known about poverty in rural India as a result of prior research, the world is changing and with it the situation of the poor. In this research initiative, the research team was able to gain new information, in part because the circumstances and conditions surrounding Indian poverty have changed since some of the earlier studies were conducted. In addition, new insights emerged from this study because the integrated research approach yielded a more comprehensive view of the multifaceted phenomenon of poverty in rural India.

The second objective was to provide a better understanding of the nature of poverty. The study was designed with the explicit recognition that poverty is an extremely complex phenomenon. This is particularly true in rural India, where the poor face not only economic constraints but social and cultural constraints as well. To gain an understanding of questions of mobility and vulnerability, it is essential to explore the social and cultural context in which poor households attempt to improve their well being and avoid further impoverishment. However, traditional survey-based methods, taken alone, are not particularly well designed to address these issues. In pursuit of this objective, therefore, the research team used a combination of methods to examine the ways in which Indian poverty is affected by its sociocultural, as well as its local political, context.

The third objective underlying this study was to provide information that could improve operational planning and development. In order to develop the most effective country program possible, it is essential that the World Bank gain a more in-depth understanding of the impacts of India's antipoverty programs, which are large and very costly. A great deal of evaluation research on these programs has been conducted in the past. However, many of those studies took years to complete, and the lessons learned were not always reflected in operational planning. Therefore, the researchers were careful to involve operational staff in the design and implementation of this study from the outset.

Approaches and Assumptions

The approach of this study was participatory, adaptable, and focused on learning. It was not intended as a definitive model of how to do integrated research. Rather, it should be considered as one example, from which the research team learned a great deal. It is expected that other teams will develop their own models, and it is hoped that they will share what they learn so that a greater understanding of integrated methodologies can emerge and evolve over time.

The researchers' basic assumption was that different methods of data collection are designed to answer different kinds of questions. On the quantitative side, a survey attempts to investigate certain variables that are measurable. Survey methods are well designed for statistical representation and for establishing central tendencies. Qualitative data, by comparison, are frequently collected and reported in the form of textual data that rely on context. In some cases, the data will report the respondents' own perceptions and aspirations, while in other cases they reflect the researchers' impressions and interpretations of the situation. In yet other cases, the researcher may record what she hears and observes in terms of categories defined in the research protocol. Frequently, the researcher wants the respondent to frame the issues or to give a highly nuanced view of events, circumstances, and social codes.

At the outset of this study, the researchers kept open the options to quantify data obtained through qualitative research and to ask open-ended survey questions. In the end, however, it proved to be more productive to use the different methods to gather different kinds of information. Experience has established for this team that when attempts are made to simplify qualitative information for quantification, its unique insights are often lost. Similarly, attempts to add *why* questions to the survey yielded responses that were too general or superficial to be useful in explaining the phenomenon in question. While it is technically possible for interviewers to record comments verbatim using standard probes, and then to use coders to classify answers post hoc, the process is costly and does not always yield useful data.

Implementation of the Research

The organization of the study was flexible and iterative, in that the approach and research tools evolved throughout the study period in response to what was learned at each stage. For example, the qualitative instruments were modified twice after the field teams pretested them and then met to share their results and exchange ideas and suggestions. The quantitative instrument was developed during and after a three-day seminar in which the field team leaders reported on and discussed the qualitative findings and analysis.

The research was conducted in two phases. The first phase consisted of qualitative research, using rapid rural appraisal (RRA) and participatory rural appraisal (PRA) methods. Because the researchers felt it was important to maintain a consistent methodological approach across the sample villages, six research instruments were prepared and a general protocol for their use was developed. The composition of instruments reflected the interests of the India country program as well as those of the local research teams. The instruments guided the teams in conducting exercises that included social mapping, wealth ranking, an assessment of services and programs, an inventory of social capital, and an exploration of gender roles, in particular gender violence.[3] The final exercise was a series of case studies. All of the exercises were carried out in each village, but the field team leaders were given authority to make adaptations to suit the specific conditions they encountered at each location.

The five local team leaders participated in pretesting and revision of the instruments, and they received training in the use of all the instruments. Designing an appropriate training program was a challenge, since some of the team leaders had a strong qualitative background but others were primarily economic researchers. This challenge was met by an experienced NGO trainer, who is coordinating several projects in one of the research areas. In terms of coverage, the approach of the qualitative component could be described as intensive rather than extensive, in that each team covered six villages over a period of about three months, for a total of 30 villages.

In preparing the qualitative phase, the primary researcher spent approximately six months in the field working with the team leaders, forming the teams, and discussing the objectives of the study. Although such an arrangement may not always be possible, the researchers believe it was fundamentally important to the success of the work. The qualitative instruments changed in some fundamental ways over the course of these meetings, reflecting local knowledge, local skills, and the process of clarifying and refining the issues.

Toward the end of the qualitative phase, the researchers met to design the survey questionnaire for the second phase of the study. Results of the qualitative research were drawn into the process of developing the survey instrument. Although the primary purpose of the qualitative work was to provide an information base and analysis that are valuable in their own right, that phase of the project was also useful in improving the quality of the questionnaire, as well as in informing, validating, and adding explanatory power to the survey analysis overall.

The questionnaire utilized the framework of the Living Standards Measurement Survey (LSMS)[4] developed by the World Bank, which measures income, consumption, and a wide range of other variables. Extra sections were added to reflect some of the findings of the qualitative study. For example, the qualitative results suggested that vulnerability to risk is a significant consideration among the poor in rural India. Consequently, the researchers added a "vulnerability" section that included questions about issues that were frequently cited in the PRA work, such as access to programs and services and reliability of food supplies. Caste and social class were also found to play an important role in influencing the distribution of opportunities and constraints. In addition, the qualitative results revealed a great concern with employment opportunities, income-generating opportunities, and stability of employment. These issues were mentioned repeatedly in the focus group discussions about poverty and risk as well as in discussions about mobility and change—especially in the context of perceptions about the possibilities for escaping poverty. An extensive employment section, aimed at measuring various on-farm and off-farm income-generating activities and the stability of those employment activities over the year, was therefore included in the questionnaire development phase.

One of the most productive exercises conducted in the qualitative phase was the wealth ranking exercise. This entailed a random drawing of 35 to 40 households in the village and asking focus groups from both poor neighborhoods and wealthier neighborhoods to rank the households with respect to wealth. During the quantitative phase, the field teams returned to those villages and surveyed the households that had been ranked. The data obtained through the survey then provided an objective measurement of the level of poverty in these households, as measured by consumption. The quantitative results are in the process of being compared with the wealth rankings obtained through the qualitative research to determine the extent to which they differ.

Selection of Research Sites

There was a great deal of discussion among the researchers as to how the sample villages should be selected and whether a random sample should be used. After extensive debate, approximately 120 villages in eastern Uttar Pradesh and north-central Bihar were identified for the quantitative survey. Ninety were chosen at random using a statistically acceptable procedure. The remaining 30 villages were chosen because they had participated in the qualitative research component, which meant that quantitative data from these villages could be used to triangulate the results. The 30 villages for the qualitative study were not chosen at random and were never intended to be statistically representative of the population. Rather, they were selected because they reflected certain characteristics that were of interest to the researchers (e.g., they were somewhat isolated yet large enough to have a certain level of caste complexity). To a certain extent, this process was similar to theoretical sampling, by which a sample is selected with a view to developing a theory that has generality. However, because these villages also participated in the quantitative survey, it is also possible to place them in the context of the wider sample and to compare them with villages with similar socioeconomic characteristics.

Because of the rigors of statistical sampling, the researchers did not overlap the samples beyond the village level by including the same households in both the quantitative and the qualitative research. However, as described above, the households that were wealth ranked were included in the quantitative survey in order to obtain a more objective indication of their economic status.

Preliminary Findings

The early findings from both the qualitative and the survey-based work may challenge some prior assumptions of the World Bank's India country program staff. This is particularly true with respect to the effectiveness of various antipoverty programs, which Bank staff thought they understood fairly well. For example, certain employment-generating schemes have been perceived by most of the India country program staff as quite successful, whereas a major food subsidy program was thought to be much less effective due to the expense and inefficiency of the public distribution system. However, both the quantitative and the qualitative research showed that the food subsidy program was the one intervention that seemed to be making a difference in the lives of the poor. The employment-generating schemes, by comparison, were found to be essentially dysfunctional, with only two percent of the sample having obtained employment through any of the programs. Another initiative that has been perceived as relatively efficient is the Integrated Child Development Services (ICDS) program. In the study areas, however, that program was found to be essentially dysfunctional. In the few villages where the program was found to be operating, the children using it were generally not poor. Interviews with officials at the state level and a review of the supply numbers had indicated that the employment programs and the ICDS program were functioning well. However, the results of this research suggest that although money is being spent on these programs, services are seldom delivered. This finding was consistent across the qualitative and the quantitative research.

The multiple-methods approach was found to be particularly productive in this context. Since the issue of the usefulness of government programs to the poor is particularly germane to the operational goals of the study, an intensive, three-phased approach was used to examine this question. First, qualitative discussions revealed villagers' perceptions of each program in terms of how useful it is, who benefits, and how program benefits are actually utilized to meet daily and/or crisis needs. The sample survey then validated these insights statistically, and analysis of the quantitative data allowed the researchers to establish statistical associations between program usage and income levels, as well as other factors such as caste identity. Finally, a follow-on evaluative substudy of the Targeted Public Distribution System (TPDS) was undertaken in Uttar Pradesh to confirm and further elucidate the findings. The TPDS substudy itself used an integrated approach. It included a more detailed analysis of the quantitative data, complemented by field research in the Uttar Pradesh villages, which utilized semistructured interviews and focus group discussions to explore the operation of the TPDS system at the local level.

The researchers believe the use of various methods in combination increases the legitimacy of the findings, at least until they are explored further. Certainly there is tremendous regional variation in India, and this study cannot claim to describe the country as a whole. However, the regions that were studied are among the poorest in the country, and they have a population of roughly 140 to 160 million. Consequently, while the findings are surprising, they are probably quite plausible considering the extreme poverty of these regions.

Another issue that emerged on the operational side is related to the move toward decentralization in India, which has shifted control of many programs to the local level. The research revealed that if the rural elites obtain control of those programs, there is a strong risk that their benefits will become part of the traditional political distribution system. These results suggest that, while there are many potential benefits of decentralization, it could actually exacerbate long-standing inequalities if it is not carefully applied.

Finally, the qualitative work revealed the central importance of social class and caste in influencing access to resources, benefits, and social capital in the communities that were studied. Although this finding may seem self-evident in the context of India, it could nevertheless have been missed if a conventional survey alone had been used. Although the existence of caste and the stigma associated with low caste identity are common knowledge, caste is generally considered to be a social rather than economic phenomenon. Without the open-ended, probing approach used in qualitative research, the many complex and subtle ways in which caste standing influences economic behavior and opportunity might never have been uncovered or included in the survey questionnaire.

Through the social mapping exercise, for example, it became evident that virtually all of the resources in most communities were clustered around the wealthier, upper caste areas. Moreover, interviews with lower caste people regarding the issue of mobility—who gets ahead in the community, and how—clearly showed that it is difficult for the respondents to imagine the possibility of moving outside the class into which they were born.

The approach of the scheduled caste and tribe informants to improving their lot in life seldom entailed striving for those assets they described as indicative of wealth and status. The lowest-caste respondents did not aspire to acquiring the characteristics of the upper castes, such as extensive land ownership, possibly because they consider this a nonviable option within their society. Nor did the majority express the hope of higher wages in their usual occupation—working on someone else's farm. Although some respondents indicated a desire for a more secure situation in their own village, most said they want stable employment outside of the village.

Respondents also indicated that they believe it is risky to leave the village. Although the village and traditional employers may not be perceived as benign, even the lower castes can expect certain protections from members of the upper castes. These protections may include various types of assistance in times of emergency and access to credit during periods of acute shortfall. When they leave the village and seek employment outside the agricultural sector, the poor lose those safeguards.

This finding suggests an innovative way of viewing antipoverty programs. Rather than just providing for minimum subsistence, these programs can also help to cushion the household against risk, thereby promoting economic behavior that is riskier but may yield higher incomes. For example, having a mechanism in place to provide for subsistence needs in the village may make it more feasible for a worker to leave his or her household temporarily and find more remunerative work (compared to casual labor in the agriculture sector) outside the village. In addition, the Government of India's antipoverty programs have the potential to replace traditional systems of support or security within the village. The study suggests that many of the traditional, vertical relationships that have helped to ensure security in the past are highly exploitative and cannot always be depended on in times of need. Thus, in the case of Uttar Pradesh and Bihar, it may be a very good thing if formal social security mechanisms (primarily government) were to crowd out informal systems.

The qualitative research also provided important information about social capital in the communities that were studied. It was learned that, first, the poor seem to recognize and maintain fewer potentially helpful social and kinship connections than do the nonpoor, second, the mutual assistance networks of the poor tend to be restricted to the immediate community and family and, third, the poor appear to use those connections primarily to defend themselves and to reduce risk and vulnerability. By comparison, wealthier households possess social ties that extend outside the village and that can be used to help family members obtain lucrative employment or secure investment capital.

Lessons Learned

The first and most important lesson of this study is that there is significant value added by using both qualitative and quantitative methods to think about and learn about poverty. It is clear that new insights have been gained that are unlikely to have been obtained from the use of either method alone. The issue that must now be addressed by the quantitative researcher is how to maximize these synergies in order to be able to incorporate the findings within the statistical framework of the study.

Second, the combined approach can make a significant contribution to the process of interdisciplinary learning. In the context of this study, the integrated approach provided a forum for the exchange of ideas between nongovernmental organizations in India (many of which had carried out poverty-related research on their own), academic researchers specializing in poverty studies at Indian universities, and World Bank staff.

Conclusions and Recommendations for Future Research

Although the combined qualitative-quantitative methodology has proved to be productive, the most useful sequence of components is still to be identified. This study led off with the qualitative component because the researchers intended to use its findings to more accurately tailor the survey questionnaire to the local context. Qualitative research may also be utilized as a follow-up to sample surveys; however, in that context discussion and probing is sometimes needed to explain puzzling or counterintuitive survey results.

In an ideal world, both options could be chosen. For example, after completing the qualitative and quantitative components of a study such as this one, it may be useful to return to the study villages and conduct more qualitative research to clarify any issues that the statistical analysis left ambiguous. However, a second qualitative phase would significantly increase the amount of time needed to complete the overall assessment, and the exigencies of operational work schedules may preclude a second round if the study is to be useful to ongoing program operations. Since choices must be made, it is recommended that future integrated research projects attempt to establish the most effective and fruitful sequencing of qualitative and quantitative exercises to best capture the advantages of the combined approach.

The findings of the integrated study also raised further questions for future research on poverty in India. For example, findings on health and education indicated that community-level schools and health care services provided by government are underutilized by the poor in the research area. Further inquiries should be initiated to fully explain these results and to identify the requirements of human services systems that would more effectively satisfy perceived health and educational needs. The results of this study also suggested that poverty is not uniform and that different types of poverty are found in the study area. This premise should be further examined so as to clarify the unique set of constraints and opportunities faced by each category of the poor, and to identify the best ways of tailoring programs and services to their needs.

In the context of the India poverty assessments, the researchers suggest that the trend toward integrated, phased, and iterative assessment activities should be encouraged. Poverty assessment is now seen as a continuous process with different phases and types of assessment, each with its own contribution to the full understanding of poverty. This is very different from the traditional, one-time assessment that was thought to be finished once the researchers had presented the findings. The new approach, which is consistent with this study's sequential use of a combination of research methods, may create new opportunities to improve coordination with operational staff so that the results of poverty research are better represented in the development of program operations. It may also allow more time and opportunity for Bank staff to better work with, listen to, and learn from their host-country counterparts.

Valerie Kozel is a senior economist in the Poverty Reduction and Economic Management Unit, South Asia Region. Barbara Parker is an anthropologist and consultant.

Notes

1. Based on Kozel (1998).

2. From the Bank side, the work was overseen by Valerie Kozel, SASPR, with additional inputs from Barbara Parker (anthropologist, consultant), Giovanna Prennushi, PRMPO, Peter Lanjouw, DECRG, and Salman Zaidi, DECRG. The study team in India was headed by Professor Ravi Srivastava (Department of Economics, Allahabad University). Field team leaders, who prepared the background papers on which much of this paper is based, include Professor Nisha Srivastava (Department of Economics, Allahabad University), Madhavi Kuckreja (Vanangana, Karvi, Uttar Pradesh), Ajay Kumar (Center for Action Research and Development Initiatives, Patna, Bihar), Sandeep Khare (Vigyan, Lucknow, Uttar Pradesh), and Sashi Bhushan (Patna, Bihar).

3. For a summary of participatory methods, see Rietbergen-McCracken and Narayan (1997).

4. For an overview of the LSMS survey program, see Deaton (1997).

References

Deaton, Angus. 1997. *The Analysis of Household Surveys: A Microeconomic Approach to Development Policy.* Washington, D.C.: The World Bank.

Kozel, Valerie. 1998. "Social and Economic Determinants of Poverty in India's Poorest Regions: Qualitative and Quantitative Assessments." India Country Program, Washington, D.C.: The World Bank.

Rietbergen-McCracken, Jennifer, and Deepa Narayan, eds. 1997. *A Resource Kit for Participation and Social Assessment.* Social Policy and Resettlement Division, Environment Department, Washington, D.C.: The World Bank.

5
Studying Interhousehold Transfers and Survival Strategies of the Poor in Cartagena, Colombia[1]

Gwyn Wansbrough, Debra Jones, and Christina Kappaz

This chapter presents experience using a combination of quantitative and qualitative research methods in a study of interhousehold transfers in the Southeastern Zone (SEZ) of Cartagena, Colombia, conducted as a follow-up to World Bank research on this topic conducted in the SEZ in 1982. The authors discuss the design and implementation of the study, which included both quantitative and qualitative approaches, and present their observations regarding integrated research. Issues addressed include survey design, sampling techniques, selection and training of interviewers, and timing and cost of research activities.

Background on the Study and Research Methodology

Very few poor families could survive over a long period of time on their resources alone. In order to supplement income, households rely heavily on transfers of money, goods, and services from friends, relatives, and neighbors. Measuring poverty and attempts to understand survival strategies of low-income households present an interesting case for integrating quantitative and qualitative research methodologies. While the existence of transfer networks can be readily identified, exact measurements of the extent, composition, and dynamics of those networks has proven to be extremely difficult to obtain.

The importance of interhousehold transfers as an essential survival strategy for poor families has been discussed for many years within the social science literature, but there have been few systematic efforts to quantify these transfers. The quantitative studies that have been conducted have been limited by the difficulty in measuring transfer activity. It is certainly possible to include a question on a survey asking how much the respondent receives in transfers, but it has been considered too expensive and complicated to obtain complete and reliable data on the volume and use of transfers.

The World Bank conducted one of the most comprehensive studies of interhousehold transfers in the early 1980s, in which a combination of in-depth case studies and quantitative surveys was used to examine transfers in Colombia, Kenya, and the Philippines. That work showed that for a poor, female-headed household, as much as 30 or 40 percent of its regular household income could come from these forms of transfers.[2] Therefore, this seemed to be an important area for further research.

In order to support increased research in this field, the Bank has expressed a growing interest in determining if there are cost-effective ways of using qualitative case studies and quantitative surveys together to get reasonably good estimates of the volume, flows, and uses of transfers by poor households and, particularly, female-headed households. The Bank also wanted to test whether reliable results could be obtained in a relatively short period of time

With these objectives in mind, the Bank decided in 1998 to follow up on some of the organization's original work on transfer behavior conducted in an area of low-income neighborhoods in Cartagena, Colombia, known as the South East Zone (SEZ). The SEZ is made up of some of the city's poorest neighborhoods and is marked by high levels of illiteracy, poor health, unemployment, and violence. The population of the SEZ, which began as a squatter settlement, now accounts for one-third of Cartagena's total population of 900,000.

This work was carried out by a team of six graduate students from Columbia University's School of International and Public Affairs. The 1982 research consisted of eight qualitative case studies conducted over a period of several months, accompanied by 507 quantitative survey interviews conducted over a two-month period. The 1998 follow-up study included case studies of five households that had participated in the original World Bank study, plus a quantitative survey of 160 households in the SEZ. The follow-up field research was completed over a two-week period.

The research objective of the follow-up study was to examine the volume, nature, uses, and determinants of transfers and to examine the importance of transfers for the survival and advancement of low-income households in the area. The quantitative survey was included to provide data on a sample size large enough for statistical analysis. The survey data were complemented by the information gathered through the qualitative case studies that were designed to provide a more profound understanding of transfer dynamics. The data collected through the follow-up study provide a point of comparison with the data gathered by the World Bank in 1982, making it possible to assess changes in the quantity and quality of transfers over time and their role in poverty alleviation.

This study also serves as an example of the types of information that can be obtained in a short period of time and at a relatively low cost using integrated research methodologies. This chapter will outline the methodologies used for the quantitative and qualitative components and will present issues and observations of the strengths and weaknesses of each approach and some possible benefits of integrating the two.

Quantitative Research

The quantitative research entailed the design of a questionnaire, the selection of a sample, the selection and training of interviewers, and the collection of data. Prior to arriving in the field, the team developed a draft questionnaire based upon the survey instrument used in the original World Bank study. The original questionnaire consisted of more than a hundred detailed questions that carefully assessed the volume, size, and nature of transfers in each participating household. The revised 43-question instrument contained more general questions regarding transfers. Once in the field, the research team pretested the questionnaire on five households in the sample area to ensure the questions were easily understood. In addition, local experts from a variety of disciplines gave feedback on the instrument's content and language. Local surveyors were selected once the team was in the field. A total of eight interviewers, who worked in pairs, conducted the interviews.

Qualitative Research

The qualitative component of this study consisted of in-depth interviews with five families exploring their transfer behavior and activities. All five families had participated in the original World Bank research in 1982.[3] This allowed the case studies to illustrate the dynamic nature of transfer patterns and the changes in household transfer activity during the 18-year time lapse since the original study.

The follow-up interviews took place over a period of eight days. While this is much shorter than the two months spent on the qualitative interviews in the original study, less time was needed in the follow-up study. The families had participated in the previous study and had developed a good relationship with the original research team, which greatly facilitated the work of the second team of researchers. This cut down considerably on time that otherwise would have been required to gain the confidence of the families. Furthermore, a member of the Bank's original research team had traveled to Cartagena six months prior to the arrival of the new team and had done much of the legwork required to locate the families.

Upon arriving in Cartagena, the research team spent two days locating and making initial contact with the households with the help of local contacts familiar with the neighborhoods. The research team spent an additional two days refining interview questions and developing codes for organizing the qualitative data in a manner consistent with the survey questionnaire. The case-study methodology was similar to the one used in the original study, with open-ended questions, some probing questions to overcome the problems of recall bias, and observation of household activity. This method was developed by the team in consultation with Scott Parris, the anthropologist who conducted the qualitative research in the original study.

Two pairs of interviewers from the Columbia University team conducted the qualitative interviews. Conducting the interviews in pairs also facilitated the collection of more data in a short period of time. Implementation of five case studies simultaneously by two teams allowed for a faster learning curve, and the teams were able to identify new issues to be discussed in interviews over the eight-day period. The researchers compiled detailed case write-ups based on notes from conversations with the families during the course of the field work.

Timing and Costs

The research team spent a total of approximately 40.5 days over a period of four weeks to design and implement the qualitative and quantitative studies. This estimate does not include time spent researching the subject of interhousehold transfers and developing the qualitative approach before going into the field, or the time spent by the World Bank staff member to locate and contact the families that participated in the original research. Additional time was also spent on report preparation and data analysis after the field research was completed. Table 5.1 provides a breakdown of the time spent on each activity.

The total direct cost for the design and implementation of the *quantitative* surveys was $1,804. Each pair of local interviewers was paid $10.76 per interview. Total costs did not include any fee for the Columbia team, as the research was part of a graduate seminar. Additional time and costs related to pre- and postfieldwork conducted by the team of six graduate students is not included in table 5.1. Two students dedicated approximately 10–15 hours each per week to the project over a three-month period, while the other four spent about 20 hours each on the project in addition to the two weeks of full-time work by all six members while in the field. Travel, subsistence, and miscellaneous expenses related to the work of the graduate students totaled $9,000 and were financed by the World Bank and the university. The team saved additional costs by using loaned office space, and supplies were provided free of charge through local contacts.

Table 5.1. Allocation of Time by Task in the Cartagena Study

Design questionnaire	7.5
– Prepare survey (5 days)	
– Pretest (0.5 day)	
– Share with key informants (0.5 day)	
– Edit (1.5 days)	
Select study population	2
Assess and choose survey interviewers	3
Train survey interviewers	2
Debrief survey interviewers	2
Conduct 160 quantitative surveys	6
Quantitative Total	*22.5*
Tasks—Qualitative Research	
Locate households for qualitative research	3
Analyze background research on interhousehold transfers	3
Review previous case studies	2
Prepare interview methodology for qualitative research	2
Conduct qualitative research	8
Qualitative Total	*18*
Total Person Days	40.5

Issues and Observations Regarding Integrated Methodologies

The follow-up study of interhousehold transfers in Cartagena raises many important issues about the most effective use of quantitative and qualitative research methods. The study also identifies ways in which the two can be integrated in order to overcome some of the problems of relying on either method on its own.

Qualitative Research

From the research experience in Cartagena, the team found that the qualitative research revealed more about size, volume, and uses of interhousehold transfers than the quantitative research did. Qualitative tools such as in-depth interviews, observation, follow-up, and repeated contact with respondents are highly effective means of gathering information about the size and uses of transfers. These methods are also more accurate than quantitative surveys at estimating the volume of transfers, because there is a higher probability of gaining more comprehensive information about network activity that can be underreported in a quantitative survey.

There are drawbacks associated with relying too heavily on qualitative methods alone, however. The team quickly discovered that a rapidly growing, marginalized population was excluded from the qualitative case studies, because this segment of the population moved into the area between the time of the original study and the follow-up. For the sake of continuity, the families for the qualitative case studies were selected on the basis of their participation in the original study. However, by excluding the most impoverished members of the community from the case studies, the research missed an important opportunity in understanding poverty and the survival strategies of those who are most in need.

Qualitative research also tends to be time consuming and costly. As already noted, significant time was cut from the follow-up qualitative research, given the time spent in the first study to establish rapport and trust with the families, which, in turn reduced costs. Moreover, while qualitative research provides very useful detailed material, the sample is not large enough to generalize findings and establish patterns that could be applied to the larger community.

Quantitative Research

The quantitative researchers did succeed in obtaining a significant amount of data in a short period of time at relatively low cost. These data provide an interesting point of comparison to the baseline data gathered in the same community in 1982. From these data, the Bank hopes to reveal patterns about transfer activity in the SEZ and understand how these have changed over time. Key preliminary findings from the quantitative survey are summarized at the end of this chapter.

Within the time constraints, the research team managed to redesign, test, and adjust the survey instrument to increase its effectiveness. The research team also spent time locating and interviewing potential local surveyors, which aided the selection of a qualified and competent team— including community leaders, social workers, and university students— to conduct the quantitative surveys. Training and employing local researchers gave the surveyors a sense of ownership. The survey debriefing sessions became a forum for discussions about community issues and ideas on how to address community problems.

Through the process of conducting the qualitative case studies, however, the researchers became aware of a significant amount of information that was not being captured in the quantitative survey, raising questions about the reliability of the data.

This was due, in part, to the quantitative survey instrument, which limited the type of information that could be recorded. Much of the effectiveness of the original study's quantitative survey was lost in the attempt to redesign the instrument in order to gather sufficient information in a short period of time and at lower cost. In the Bank's original study, the quantitative survey was designed in cooperation with a national research institute, with an extensive list of specific questions regarding the participants' transfer behavior. Researchers with survey experience administered the surveys. The degree of detail in the surveys meant that the interviewers spent an average of two hours per household.

While the survey used in the follow-up study was based on the original survey, it was shorter in length, with fewer, more general questions regarding household transfer activity. The interviewers had some relevant training but less experience with administering surveys than the original team. Instead of the two hours spent on each survey in the original study, the interviewers in the follow-up study spent an average of 45 minutes with each household. Moreover, the original interviews were conducted over a period of two months, whereas the follow-up surveys were administered over a period of ten days. Finally, while approximately 500 interviews were administered in the first study, only 160 were administered in the follow-up study, resulting in fewer cases upon which to establish the baseline data.

Through the process of conducting the surveys, the researchers discovered ways of shortening the survey without sacrificing its effectiveness, but these were not incorporated into the instrument. The survey instrument could have been more effective if it had focused on one or two themes within transfer activity instead of attempting to capture information about all aspects of transfers. To capture more complete information, the survey could also have included more effective probes and recall mechanisms to replace the original survey's extensive questions on transfer activity. In retrospect, it would have been possible to design a longer survey and still complete the quantitative component within the allotted time period and cost constraints.

Beyond the problems associated with the redesign of the survey instrument, the very nature of the subject matter does not always lend itself to the use of the survey method. It can be difficult to get an accurate picture of the size of the transfer network with a survey. Transfer networks consist of informal relationships among close family, friends, and neighbors. The qualitative researchers found that with each conversation with the case-study families, the reported size of the network expanded, the transfers grew in value, and relatives who had not previously been mentioned suddenly became very important givers or receivers of transfers. This new information drastically altered the researchers' initial impressions of the family's transfer activity, providing a much more complete picture of the resources on which households can draw.

In addition, a survey relies on the respondent's ability to accurately report his or her behavior. However, people are not always conscious of activities, such as transfers, that are an integral part of their lives. Participants in interhousehold exchanges do not generally view themselves as part of a "transfer network" and may not recognize something as a transfer that the researcher would classify as such.

Quantitative surveys are limited in their ability to capture sensitive information. Transfer activity, for example, touches on people's relationships, their finances, their pride, and their dependency on others to supplement their income. It often takes a certain amount of time and repeated contact to establish a level of trust and rapport that enables a respondent to disclose such information. Qualitative methods, including observation, conversation with many members of a household, and interaction over an extended period with the respondent are better suited to drawing out difficult-to-reach information due to problems of recall and/or sensitivity.

Another problem in using a quantitative approach to study transfer activity relates to the difficulty of assigning a numerical value to noncash transfers. In order to estimate the volume of transfers or their relative importance to household income, one must be able to estimate the monetary value of noncash transfers that take the form of goods, services, people, and information. This is especially relevant for women, who are perhaps most often in a position to exchange services, such as collective childcare, whose value is not easily quantified.

It is also difficult to estimate the quantitative value of a favor, yet this can be an important form of transfer. In one case observed by the qualitative research team, a favor extended to the son of a housekeeper greatly assisted him in finding a well-paid job, which in turn enabled him to leave the low-income neighborhood and enhance his socioeconomic status. This intangible transfer of an opportunity to increase one's ability to generate future income is important to understanding how people gain access to other transfer networks with more resources than their own and improve their own standard of living. If the value of these transfers cannot be estimated, and if they are not included in the quantitative surveys, then the quantitative research will fail to capture a wide range of transfers that are extremely important to the dynamics of the community.

Integrated Methods

While this study highlights some drawbacks of each approach, it also demonstrates that there are a variety of ways of integrating quantitative and qualitative methods to achieve higher quality research results.

First, findings from qualitative research can inform quantitative research design. Due to time constraints, the quantitative and qualitative components of this study were conducted simultaneously. In retrospect, the researchers believe that the two methods may have been more complementary if they had been conducted sequentially. If the qualitative case studies had been conducted first, the information gained through those interviews could have been used to design a more effective survey. The researchers would have had a better understanding of the range of issues, the kinds of questions that should be asked, the probes that might be used to access that information, and how to more effectively adapt the survey instrument to the local context.

Second, qualitative methods could be used in quantitative research. A flexible quantitative survey format that allows for dialogue between the surveyor and respondent would have enabled the surveyor to draw out information about transfer activity that is sensitive or difficult for the respondent to recall.

Finally, this research suggests that the relative importance of qualitative and quantitative data needs to be reexamined. Qualitative research is frequently used to verify findings from quantitative research. Qualitative case studies are also used to give a "face" to numbers by providing people's experience to accompany the baseline data. In both cases, the qualitative research plays a supporting (or subordinate) role to the quantitative findings. This study suggests that more accurate information was gathered through qualitative methods; therefore, the quantitative findings should be used to back up the qualitative research and not the other way around.

In conclusion, the study of interhousehold transfers demonstrates that qualitative and quantitative approaches each increase our understanding of interhousehold transfers and poverty. If designed properly, qualitative research can provide data that point to general patterns and tendencies in transfer behavior. While qualitative research can be more costly and time consuming, it can provide reliable and in-depth information about the dynamics of transfer activity. However, each approach on its own gives us only a partial understanding of transfer activity. This study serves to illustrate the benefits of using integrated approaches in social science research and outlines some ways of combining qualitative and quantitative methods using the example of a study of interhousehold transfers. It is important that researchers understand the constraints of each approach, try to find the complementary aspect of the other approach, and use the two in tandem.

Preliminary Research Findings

Despite some of the weaknesses of the quantitative survey process discussed above, the results of the 160 surveys conducted in the SEZ do provide some important illustrations of transfer behavior which reflect many of the patterns identified through other studies. While it has proven to be difficult to identify strong determinants of transfer behavior, findings from several case studies point to general patterns of transfer giving and receiving based on income and gender.

Income and Transfer Behavior

Income level appears to be an important determinant of whether a person will be a transfer giver or receiver. Transfers generally flow from households that are better off to those that are relatively worse off. The Cartagena case study supports this general finding, showing that transfer giving increases steadily with income. Table 5.2 summarizes the proportion of households surveyed who give and receive transfers, according to income quintiles. The data reveal that half of the households in the highest income quintile are considered to be "transfer givers," as compared to only 13 percent of households in the lowest income quintile.

It follows that low-income households are more likely to be "transfer receivers," although in different proportions than those observed with transfer givers. In the two lowest income quintiles, approximately 22 percent of households are considered to be net transfer receivers, while in the highest income quintile only 3.3 percent are considered to fall into this category. Unlike transfer giving, which rises steadily with income, the proportion of transfer-receiving households rises sharply in the second-highest income quintile from 6.7 percent to 16.7 percent and falls again in the highest income quintile. The increase in transfer receiving households in the second-highest income quintile may be explained by efforts of individuals in this income bracket to jump to the next income bracket by accruing debts and generally relying on assistance from others to do so.

Table 5.2. Proportion of Households Giving or Receiving Transfers by Income in Quintiles (percent)

Income quintiles	Neither given nor received	Giving	Receiving	Both give and receive
0–240,000	38.71	12.90	22.58	25.81
240,000–370,000	57.14	17.86	21.43	3.57
370,000–600,000	36.67	33.33	6.67	23.33
600,000–1,000,000	30.00	36.67	16.67	16.67
> 1,000,000	26.67	50.00	3.33	20.00

Source: Cartegena SEZ Quantitative Survey 1998 (World Bank, unpublished).

It is difficult to draw a general pattern from the proportion of households that both give and receive. One point that was raised in the qualitative case studies is that in order to maintain network ties it is important that the relationship be reciprocal, even if the exchange is unequal in value. Thus, it would fit that the highest proportion of households that give and receive is in the lowest income quintile (26 percent), where transfer networks are generally more important for survival.

The data from the Cartagena research are inconclusive regarding the nonparticipation of certain households in any type of transfer network. The proportion of households not participating in transfer networks appears to be higher than would have been expected based on studies that suggest that participation in transfer networks is widespread, and that transfers are particularly important to the survival of lower income households. The results of the quantitative study indicate that for the three lowest income quintiles, a significant number of households neither give nor receive transfers. In the case of the second-lowest income quintile, 57 percent of households claim not to give or receive transfers. This appears to run contrary to the assumption that households with lower incomes are the most reliant on transfer networks for survival. There are several possible explanations for this discrepancy. One probable explanation is that the barriers to reliable quantitative data collection on the subject of transfers which were discussed above could have contributed to a larger proportion of respondents appearing to be transfer-avoiders than is actually the case. Findings from the qualitative case studies support this explanation in that they suggest that most families participate in some way in transfer networks.

However, the qualitative case studies also suggest an alternative explanation since they demonstrated that the participants would prefer independence from transfer networks. Participants in the qualitative case studies explained that while participation in transfer networks is widespread, and in some cases important to household survival, transfers are unreliable and insufficient to meet the needs of the household.

Another interesting finding from the quantitative data is that the highest income quintile represents the lowest proportion of households who neither give nor receive transfers. This is consistent with the findings, indicated above, that households in the highest income quintile account for half of transfer-giving households and are therefore tied into the transfer networks.

Gender and Transfer Giving and Receiving

Research on transfer patterns has shown that female-headed households are more likely to be recipients rather than givers of transfers. The results of the quantitative study (Table 5.3) appear to support the general finding that a higher proportion of male-headed households are classified as transfer givers than female-headed households. Contrary to the general finding, however, the results of the study also show that a slightly higher proportion of male-headed households are transfer receivers. It is possible that some of the female-headed households that are net transfer receivers fall into the category of households that both give and receive, given that female-headed households account for a higher proportion of households that both give and receive.

The proportion of female-headed households that neither give nor receive transfers is higher than that for males. This is also surprising, given that general findings on gender and transfers point to the reliance of female-headed households on transfer networks. That female-headed households in this study represent a higher proportion of households that do not participate in transfer networks may not be by choice of the female head. Often network ties, tied to the male partner, are cut when the male partner leaves the household. The discrepancy could also be explained, as in the case of income above, by the possibly high rate of error in the quantitative data. The issue of transfer receipts could be particularly sensitive to female-headed households who, for various reasons, could have assumed it was in their best interests not to reveal to an unknown interviewer assistance received by others. These reasons could include the sensitive nature of personal relationships or a perception that the survey results might be used in determining public assistance programs, in which case the respondents would have wanted to emphasize their precarious economic condition.

Table 5.3. Proportion of Households Giving or Receiving Transfers by Gender of Head of Household (percent)

Gender of head of household	Neither given nor received	Giving	Receiving	Both give and receive
Male	31.58	36.84	15.79	15.79
Female	44.44	20.63	14.29	20.63

Source: Cartegena SEZ Quantitative Survey 1998 (World Bank, unpublished).

At the time this chapter was written, the authors were graduate students at the School of International and Policy Affairs, Columbia University.

Notes

1. This chapter is based on a report written by Gwyn Wansbrough, Christina Kappaz, Lauren Bergner, Lissette Bernal Verbel, Debra Jones, and Provash Budden in May 1998 as part of a cooperative program between the School of International and Public Affairs, Columbia University, and the World Bank.

2. Bamberger, Kaufmann, and Velez (1997).

3. Parris (1984).

References

Bamberger, Michael, Daniel Kaufmann, and Eduardo Velez. 1997. *Interhousehold Transfers and Survival Strategies of Low-Income Households: Experiences from Latin America, Africa, and Asia.* Washington, D.C.: The World Bank. Mimeo.

Parris, Scott. 1984. *Survival Strategies and Support Networks: An Anthropological Perspective.* Washington, D.C.: The World Bank.

Education

6
Evaluating Nicaragua's School-Based Management Reform

Laura B. Rawlings

This chapter discusses how a mixed-method approach combining quantitative and qualitative research was used to evaluate the impact of Nicaragua's school decentralization reform. Following an overview of the reform program and the objectives of the evaluation, the author discusses the quantitative and qualitative techniques utilized in the study, sampling issues, and the sequence of activities. Drawing on the results of the full research project, the chapter summarizes the findings of the first phase of the study regarding the role of the school in governance, the perceived level of influence of key stakeholders (directors, teachers, and school council members), and the perceived impact of the reform on school performance. This chapter contributes to the evaluation of the Nicaraguan reform by clarifying the objectives, methods, and value-added of the mixed-method approach.

Purpose and Background of the Evaluation

In developed and developing countries alike, decentralization is being actively employed as a means for improving the quality and effectiveness of public services. In Latin America, decentralization has often accompanied democratization in an effort to make public institutions more directly accountable to local stakeholders. In the education sector, school-based management reforms bring decision-making directly to the school, calling upon parents, school staff, and local stakeholders to assume a central role in the school's organization and management. The theory behind this type of decentralization reform is that schools will become more efficient in managing increasingly scarce resources and more effective in instructing students.

Despite its growing popularity, school-based management is seldom evaluated systematically with respect to its impact on schools.[1] This study was part of a major effort to do so in Nicaragua, which launched a school autonomy reform program in the early 1990s aimed at improving the quality and efficiency of the primary and secondary public education system. The evaluation was initiated in 1994 and includes a longitudinal school-household survey, an in-depth qualitative process evaluation, and student achievement tests. This chapter discusses the first phase of the evaluation, which is based on the results of the first school-household survey and qualitative study and is oriented toward assessing process changes within schools. The second phase of the research that is presently being carried out is analyzing the results of the longitudinal school-household survey and the achievement tests to assess the impact of the reform on student performance.

The evaluation effort has been led by the Nicaragua Reform Evaluation Team, which consists of staff from the Nicaraguan Ministry of Education, the World Bank, and U.S.-based university researchers. Four working papers have been produced to date on the results of the evaluation effort.[2]

The principal questions addressed in the first phase of this study and covered by these three papers were:

- whether or not autonomous public schools exercise greater control over their management than do traditional public schools;
- whether or not (and which) local stakeholders (directors, teachers, council members) affect school decisions; and
- how local stakeholders perceive the changes that have occurred in schools since autonomy.

The Reform Program

The decentralization reform evaluated through this research is a principal element of the education policy of the coalition government that replaced the Sandinista regime in 1990. Compared to other educational decentralization reforms on the basis of the degree of authority transferred to schools, Nicaragua's policy represents one of the most profound policy reforms initiated worldwide.[3]

The Nicaraguan reform aimed to give schools power over key managerial and pedagogical decisions and transfer financial administration directly to the schools. The reform program began in 1991, when the Ministry of Education established councils (Consejos Consultivos) in all public schools to facilitate community participation in decision-making. The councils are composed of the school principal, teachers, parents, and students.[4] Except for the nonvoting students, each member has an equal vote, with a simple majority required to reach decisions in all areas under the council's jurisdiction.

The program was expanded in 1993 through a pilot program that transformed the councils of 20 public secondary schools into school management boards (Consejos Directivos), thereby creating "autonomous" public schools. Key management tasks were transferred from central authorities to the directive councils, with the objective of increasing the efficiency of use of public funds, mobilizing local resources, and improving school effectiveness. By the end of 1995, well over 100 secondary schools had signed a contract with the Ministry of Education to establish a directive council and become autonomous.

In 1995 the reform was extended to primary schools, at which time school autonomy took on two forms. Directive councils similar to those in secondary schools were established in urban primary schools, while a new model—the Nucleos Educativos Rurales Autonomos (NER)—was introduced for rural schools. Each NER consists of two to four schools, formed around one center school, that act as one autonomous school with a shared council. The NER's directive council is based in the center school, which is usually the largest in the group and the only one with a director. Today, school autonomy has been introduced in the majority of public primary and secondary schools in Nicaragua.

The public schools that have become autonomous are legally vested with many of the features of private schools. Table 6.1 compares traditional and autonomous public schools to private schools with respect to whether the responsibility for various functions resides with the Ministry of Education or the school. The differences in councils' responsibilities across schools provide a clear illustration of the transformation of Nicaragua's basic educational system and illustrate the extent to which autonomous schools are legally vested with extensive powers over pedagogy, administration, personnel, and finance. Of particular note are autonomous schools' authority over choosing their own textbooks, hiring and firing personnel, and setting and retaining students' monthly fees.

Research Sources and Methods

A combination of quantitative and qualitative approaches was purposively selected for the research in order to ensure triangulation and strengthen the analysis of the decentralization reform. In applying triangulation, the mixed-method approach sought to engage first in *corroboration* in order to ensure convergent validity from the results of each method, and second to engage in *elaboration* by using the qualitative techniques to expand upon our understanding of the reform as presented by the initial results of the quantitative data. From an operational perspective, the mixed-method approach was selected in order to build a range of evaluation capacity in the Ministry of Education.

Quantitative Methods

In our research, quantitative methods were used to generalize results for different types of schools and to assess causality through econometric

Table 6.1 Nicaragua's School Autonomy Reform: Changes in the Locus of Decision-Making Over Time

Key decisions	Prereform All public schools	Postreform Traditional public schools	Postreform Autonomous schools	Unchanged Private schools
Structuring the education system	Ministry	Ministry	Ministry	Ministry
Setting the curriculum	Ministry	Ministry	Ministry	Ministry
Formulating schools' annual pedagogical plan	Ministry	Ministry	School	School
Hiring and firing teachers	Ministry	Ministry	School	School
Promotions policy	Ministry	Ministry	Ministry	Ministry
Setting classroom hours by subject	Ministry	School	School	School
Programming extracurricular activities	Ministry	School	School	School
Selecting textbooks	Ministry	Ministry	School	School
Setting equivalencies[a]	Ministry	Ministry	School	School
Evaluating students	Ministry	Ministry	School	School
Establishing pedagogical methods	Ministry	School	School	School
Setting school fees	Ministry	Ministry	School	School

a. This pertains to requirements that must be fulfilled in order to determine the academic level of students who transfer schools.
Source: Ministry of Education, Nicaragua, 1996.

analysis. For this purpose, a matched-comparison was constructed between autonomous and nonautonomous schools. Matching between the treatment and comparison groups was based on the timing of the reform and the schools' size and location. Data limitations and the absence of a pre-reform baseline prevented the matching from being more exact. The quantitative research included the following activities:

- *School survey:* A survey was conducted of a random sample of 242 schools, including autonomous public schools and traditional public and private schools at both the primary and secondary level. The sample was representative by type of school. The survey collected data on a wide array of variables, including school enrollment, levels of student absenteeism, grade repetition and dropout, physical resources,

the education and experience of personnel, and changes within schools since the reform. Different questionnaires were developed for school directors, teachers, and council members to obtain school- and individual-level information.
 - *A special questionnaire* was developed as part of the school survey to determine whether the school made important decisions and whether the respondents felt influential in the decision-making process. The questionnaire focused on 25 key decisions, including those pertaining to academic matters (teacher training, supervising, and evaluating teachers, and pedagogy) as well as administrative decisions (salaries, school budget, and personnel).
- *Household survey:* A random sample of students was selected from each school and followed to their homes in order to obtain information on their families' socioeconomic status and parents' participation in school affairs. This took place at the same time as the school survey and included close to 3,000 households.
- *Achievement tests:* Achievement tests in math and language were applied to the sample of third grade primary and second-year secondary students in order to compare schools' academic performance across types of schools. The results of these tests are being examined in the second stage of the research.

Qualitative Methods

Qualitative research methods were used in a subsample of 18 schools to develop typologies, assess beneficiary perspectives, examine the context in which the reform was introduced, and analyze the decision-making dynamics in each school. In particular, the qualitative evaluation aimed to detect patterns and highlight variation in schools and across actors. The qualitative evaluation consisted of focus groups with parents, teachers, and school council members and key informant surveys with the school director and local official from the Ministry of Education. In all cases, informants were interviewed to elicit their views regarding their roles and expectations related to the reform in order to understand the diversity of perspectives held by local actors.

Sequence of Activities and Use of Survey Data to Select Qualitative Sample

The study was conducted sequentially in order to maximize the utility of both the quantitative and qualitative methods. The first quantitative school-household survey was conducted in November and December 1995 in the representative sample of 242 schools. The results of this survey were used to refine questions for the first qualitative evaluation as well as to select the subsample of 18 schools in which the process evaluation was conducted.

In selecting the 18 schools to be visited during the qualitative evaluation, three principles were applied. First, the stratifications applied to the quantitative sample were kept in order to ensure a balance between primary and secondary, urban and rural, and autonomous and nonautonomous schools. Then, with regard to the phenomena that interested us the most—school decision-making and stakeholder authority—we used the survey data on decision-making and levels of school activity to select schools representing extreme cases in order to clearly understand the dynamics behind these results during the subsequent qualitative process evaluation. Finally, in order to eliminate confounding influences, we made sure that our 18 schools had average characteristics for their type of school with respect to size and students' socioeconomic status.

Findings of the Evaluation

The initial evidence suggests that Nicaragua's educational reform program is successfully expanding the role of the school in governance. Autonomous schools are making significantly more decisions than traditional public schools, notably in the realm of school administration. However, stakeholders are not participating equally in the reform process. Directors have been most empowered by the reform, while teachers have largely been left out, and parental participation is mixed. From the qualitative evaluation it is quite clear that local institutions and economic constraints play a critical role in determining how successfully the reform is implemented across schools. Finally, perceptions concerning the essence of the reform and its impact are varied as well, notably across schools, but also across actors, with respect to the financial, administrative, and pedagogical elements of the reform process. These key findings of the first phase of the study are summarized below, drawing from both the quantitative and qualitative results.

Role of the School in Governance

Results of the 1995 survey reveal that at both primary and secondary levels autonomous schools do in fact exercise more control over key school decisions. Within schools, there was agreement among directors, teachers, and council members regarding the degree of autonomy their school possesses. However, there were significant differences among schools, particularly regarding the types of decisions being made and the perceptions of individual roles in decision-making. These findings draw from a scale of school autonomy that measures the number of decisions made by the school as opposed to the Ministry of Education, and the personal influence felt by school-based actors over these decisions. This scale was applied during the quantitative survey of directors, teachers, and council members. As reported in "Nicaragua's School Autonomy Reform: A First Look," the differences between autonomous and traditional public schools were most pronounced in administrative functions.[5] Autonomous schools showed more decision-making power in areas such as hiring and firing of school personnel, setting of salaries and incentives, and planning and allocation of the school budget. In many key areas of school administration, they had become more similar to private schools than to traditional schools. It should be noted, however, that decisions affecting instruction (e.g., pedagogical practices and teacher training) in autonomous schools were still controlled by the Ministry of Education, despite legal and policy attempts to transfer these functions to schools. The survey results suggest that the initial impact of the reform has therefore been more administrative than pedagogical in nature.

The qualitative evidence confirms this conclusion, adding depth and nuance. As Fuller and Rivarola (1998) report, the findings of the first phase of the research indicate that, in general, wealthier, more cohesive schools with a strong sense of mission report success with the reform.[6] Their responses highlight the effects on accountability and shared responsibility. By contrast, poorer, internally fractured schools emphasize the negative—and often financial—aspects of autonomy. These results also indicate that the local school context into which the reform is introduced has a powerful influence on the success and direction in which the reform is implemented in the school.

Individuals' Influence in Decision-Making

Using the same scale of decision-making, we next looked at the relationship between school-level decision-making and the individual influence felt by school-based actors—directors, teachers, parents, or teacher council members—over key decisions. Survey responses by directors, council members, and teachers regarding their perceived level of influence over school decisions revealed discrepancies by stakeholder. Results point to a positive association between the number of decisions made directly by the school and the level of the director's perceived influence, controlling for school and community characteristics. Council members, however, did not feel that they exert much influence over school decisions, regardless of whether the school was participating in the autonomy reform. Council members in secondary autonomous schools felt much less influential than directors, with only a slight positive relationship between the level of school-based decision-making and the perceived level of personal influence.

No relationship between school-level decision-making and self-perceived influence was found for teachers who do not sit on the council. Teachers in all types of schools at both the primary and secondary level felt they have little influence over school decisions. In fact, across all types of schools and all school-level actors, teachers reported the lowest level of involvement in decision-making, even on pedagogical matters.[7]

The qualitative data helped confirm and elucidate the survey results. When the 18 schools were visited for the qualitative evaluation, results from focus groups and key informant interviews alike revealed that the reform is known more for its administrative than pedagogical aspects.

With respect to stakeholders' participation in the reform, teachers revealed in the focus group sessions that they felt the school council did not represent their interests, and some were threatened by the role of parents on the council, who were often perceived by the teachers as being unprepared for their responsibilities. Teachers who viewed the autonomy reform as having negative effects often expressed disappointment at not having seen material gains in the form of salary bonuses materialize, notably in poorer, rural schools. Other teachers, however, perceived greater accountability and a stronger focus on student achievement as the key elements of the reform.

The qualitative evaluation also revealed mixed levels of parental participation. Rivarola and Fuller suggest that this may stem from parents' opposition to paying the school fees that many autonomous schools are requiring, combined with the nascent councils' limited capacity to involve parents more broadly. Poverty and the institutionalized view that schools are formal organizations belonging to educators and the state, not parents, may also contribute to this situation.

Effects of the Reform on School Performance

The quantitative survey interviews with directors, council members, teachers, and parents included a number of questions to assess the perceived effect of the reform on specific indicators of school performance. Respondents were asked whether student performance, teachers' attendance, and parents' participation improved, remained the same, or worsened after the school became autonomous. Overall, the findings indicated that autonomy was perceived as having a positive impact on several dimensions.

At both the primary and secondary levels, about half of the respondents in autonomous schools felt that students' academic performance improved as a result of the reforms; most of the rest feel it was unchanged. The majority of respondents reported an improvement in teachers' attendance after autonomy. There was less evidence of an increase in parents' participation.

Next we examined whether the effects of autonomy vary with time both with respect to the number of decisions made by the school versus the Ministry of Education and to distinct school-based actors' perceptions of impact. The evidence indicated that the longer the schools are in the reform, the more they adopt it, as evidenced by the positive correlation between the increase in the number of decisions being made directly by the schools, and the schools' years of participation in the reform. School-based actors' survey responses suggest that directors were the first group to feel the benefits of the reform. Comparing recently autonomous secondary schools with those that were early reformers, the directors' responses indicated that the longer the duration of autonomy, the better they felt about the reform and its accompanying changes. This pattern did not emerge, however, for teachers, council members, or parents, whose perceptions were not affected by how long the school had been autonomous. Moreover, the pattern of directors' responses was not as clear for schools that had been autonomous for less than two years. It may take more than two years for the majority of directors to see improvements, or the reform may inherently yield the greatest potential benefits to larger schools, which were the first to become autonomous. Finally, because the reform was introduced in hand-picked, large urban schools, the implementation process itself may be affecting the survey results. A more considered judgment of the reform's impact on student and school performance will have to await the results of the second phase of the research.

Several of the autonomous schools included in the qualitative evaluation are focused on improving school performance, mainly through actively monitoring student attendance and dropout rates. Few autonomous schools, however, had made a link between the reform and improving instructional leadership. Instead, Fuller and Rivarola report, several directors in the sampled schools have become "petit bureaucrats" focused on the administrative aspects of the reform, but failing to mobilize their increased autonomy for pedagogical change.[8]

Both the quantitative and qualitative data revealed wide variation in school contexts, including staffing, student-teacher ratios, and poverty levels of communities that point to inequalities in the school system predating the introduction of the reform. This situation raises questions about how effectively the reform, notably the financial aspects aimed at raising the levels of resources available to the schools through increases in local revenues, can be implemented broadly, particularly in poorer communities. Indeed, in the qualitative evaluation, parents and teachers alike often voiced concerns about the financial aspects of the reform.

Impact of the Evaluation

The results of the evaluation prompted a number of changes in the school reform program. First, the role of teachers was redefined through a decision by the Ministry of Education that teachers should be elected to the council by their peers instead of being appointed on the basis of seniority. Although the Ministry of Education had considered this reform for quite some time, the evaluation was influential in crystallizing the decision. This change was perceived as a way to enhance the link between the teachers and their representation on the council. Moreover, it did not require any expenditure of funds.

Second, the evaluation led to a greater academic focus in the reform program. The Ministry of Education decided to place an emphasis on training and promotion activities with regard to the pedagogical aspects of the reform, and it is considering the establishment of a Pedagogical Council in schools. The Ministry of Education also plans to strengthen the professionalism of the regional representatives so that they can better address pedagogical issues.

Third, in response to equity and finance issues raised in the evaluation, the Ministry of Education has introduced a teacher incentive program that includes a poverty map-driven subsidy scheme, particularly for the rural areas. Nicaragua also introduced a new school-finance system aimed at achieving a more equitable distribution of resources across schools.

Fourth, the experience of conducting the study led to the adoption of an evaluation focus within the Ministry of Education. For the second stage of the evaluation, the ministry decided to focus more closely on the role of teachers as agents between the school-level reform and academic performance, in an effort to establish that link more directly.

Stage II Research

The second stage of the evaluation has also continued to include both quantitative and qualitative components. The quantitative component consisted of a second round of school-household surveys and testing in 1996–97 in order to assess program impact, following up on a cohort of students from the first round, plus surveys and testing of a new cohort of students in 1997–98. The results of this research suggest a significant, positive relationship between the level of decision-making exercised by the school and students' academic performance.[9]

The second stage of the qualitative component, which was conducted in 1998, studied a subsample of 6 of the original 18 schools that were evaluated in the first round. This round of the research focused on the link between the reforms and the teaching/learning process in the classroom and on the teacher's role as the agent between the reform and achievement. Research techniques were expanded to include some classroom observation in addition to focus groups and key informant interviews.

Value of the Mixed-Method Approach

The use of a mixed-method approach added value in a number of ways. First, it built a broad base of evaluation capacity within the Ministry of Education. The intuitive nature of the qualitative work appeared to be more accessible to Ministry of Education staff than the econometric work. The quantitative work, however, imparted robustness and generalizability to the results, aspects that were important to the decision-makers who initiated the policy reforms that were adopted in light of the evaluation. All staff, policymakers, and researchers appreciated the sequential approach to the research that allowed the quantitative and qualitative work to inform one another as the research unfolded and was marshalled to address specific questions.

Second, the mixed-method approach strengthened the robustness of the research results through triangulation. The qualitative work corroborated and validated the findings of the quantitative work and, by giving voice to the stakeholders, lent a perspective that would otherwise have been absent. The quantitative survey work carried out in a representative sample of schools allowed the results to be generalized across types of schools, an element that would not have been possible using the qualitative work alone. It also laid the foundation for an impact evaluation to be carried out in the second stage of the evaluation.

Third, the policymakers drew upon the paradigms that were developed through the qualitative work. These provided the policymakers with a direct understanding of the individual school contexts, which may have been more difficult to convey through the presentation of quantitative data alone.

Fourth, the research provided insight into the marginalization of teachers and their critical role in the reform process. It also demonstrated to policymakers that many of the expected outcomes of the school reform program had not occurred, particularly regarding the reform's impact on pedagogy.

Finally, both the quantitative and qualitative research underscored how a school's context affects the implementation of the reform. It demonstrated that the reform alone may be insufficient to mobilize desired improvements in poor and internally fractured schools. It also demonstrated that the legal implementation of a policy does not easily translate into changes in action.

The author is a monitoring and evaluation specialist in the Latin America and Caribbean Regional Office, Human Development Division (LCSHD).

Notes

1. Summers and Johnson (1994).

2. The four papers are King, Rawlings, Özler, and the Nicaragua Reform Evaluation Team (1996); Fuller and Rivarola (1998); and King and Özler (1998) and King, Özler, and Rawlings (1999). These have been issued as working papers in the series "Impact Evaluation of Education Reforms" and are available from the Development Research Group, Poverty and Human Resources Division of the World Bank.

3. Hansen (1996); Di Gropello, and Cominetti (1998).

4. Students are included as council members only in secondary schools.

5. King, Rawlings, Özler, and the Nicaragua Reform Evaluation Team (1996).

6. King, Özler, and Rawlings (1999).

7. Fuller and Rivarola (1998).

8. Fuller and Rivarola (1998).

9. King and Özler (1998).

References

Di Gropello, Emmanuela, and Rossella Cominetti, eds. 1998. "La Descentralización de la Educación y la Salud: Un Analisis Comparativo de la Experiencia Latinoamericana." CEPAL (Comisión Económica para America Latina y el Caribe), Santiago de Chile.

Fuller, Bruce, and Magdalena Rivarola. February 1998. "Nicaragua's Experiment to Decentralize Schools: Views of Parents, Teachers, and Directors." *Impact Evaluation of Education Reforms*. Working Paper Series, Development Research Group, Poverty and Human Resources Division, Washington, D.C.: The World Bank.

Hanson, E. Mark. 1996. "Comparative Strategies and Educational Decentralization: Key Questions and Core Issues." Department of Education and Management, University of California, Riverside.

King, Elizabeth, Berk Özler, and Laura Rawlings. May 1999. "Nicaragua's School Autonomy Reform: Fact or Fiction?" Working Paper Series,

Development Research Group, Poverty and Human Resources Division, Washington, D.C.: World Bank.

King, Elizabeth, Laura Rawlings, Berk Özler, and Nicaragua Reform Evaluation Team. October 1996. "Nicaragua's School Autonomy Reform: A First Look." *Impact Evaluation of Education Reforms.* Working Paper Series, Development Research Group, Poverty and Human Resources Division, Washington, D.C.: The World Bank.

King, Elizabeth, and Berk Özler. June 1998. "What's Decentralization Got To Do With Learning? The Case of Nicaragua's School Autonomy Reform." *Impact Evaluation of Education Reforms.* Working Paper Series, Development Research Group, Poverty and Human Resources Division, Washington, D.C.: The World Bank.

Summers, A., and Johnson, A. 1994. "A Review of the Evidence of the Effects of School-Based Management Plans." *Review of Educational Research.*

7
Evaluating the Impacts of Decentralization and Community Participation on Educational Quality and the Participation of Girls in Pakistan

Guilherme Sedlacek and Pamela Hunte

This chapter focuses on the qualitative component of mixed-method research conducted in Pakistan to evaluate the impacts of decentralization and community participation on the quality of schools and to identify factors that influence the educational participation of girls. The authors discuss issues such as research design, sampling methods, and research questions and present preliminary findings and their implications for the project. The authors also suggest a model for conducting follow-on studies in consultation with the community and offer recommendations for conducting integrated research.

Research Objective

Although Pakistan has a significant number of schools and teachers, many parents do not send their children to school. The World Bank prepared several projects to address that problem, including the Northwest Frontier Primary Education Project and, simultaneously, sector work to improve basic education in Pakistan.

This research was conducted in an attempt to understand how the enrollment of girls in primary schools in Pakistan can be increased, how absenteeism and dropout rates for girls can be decreased, and what policies should be in place to improve the quality of girls' education.

Types of Studies Conducted

As background for the sector work, five studies were conducted to identify the constraints or disincentives that families face when sending their children to school. The studies included the following:

- A comparison of the private and public educational systems to determine the constraints that led parents to reject the public system.
- A study in Balochistan, conducted as part of an experimental design to deliver schooling through government and through partnerships with government.
- A survey of schools, teachers, and households in the Northwest Frontier to understand patterns of absenteeism in terms of incentives for teachers and students.
- An ethnographic study conducted at the community level to understand the constraints that were being analyzed quantitatively through the other studies.
- An evaluation of the Balochistan Community Support Program, which is a partnership between the government and the community to deliver schooling.

Qualitative research methodologies were utilized for the fourth and fifth studies. The first three were econometric studies.

Qualitative Research Design

The research team for the ethnographic study included Matt Miles and Ray Chesterfield, who had conducted a series of studies in Colombia, Bangladesh, and Africa that were published as "Roads to Success." The objective of these studies was to learn what constitutes success from the point of view of parents and the community, with the goal of helping the World Bank formulate policy recommendations that are consistent with local definitions of success.

Miles and Chesterfield's work was an attractive model for testing the evaluation team's belief that parents in Pakistan were rejecting the current school system. The structure of the quantitative econometric studies was not conducive to obtaining such information.

The challenge for the researchers was how to arrive at a common definition of success that would inform the World Bank's policymaking. Qualitative methodology starts from the premise that success must be defined by the community and not by the researcher. However, the World Bank has its own definition of what constitutes a successful school. This raised the question of whether there was a correlation between the schools that did not meet the Bank's definition of success and those that parents were rejecting. The researchers therefore wanted to design a study through which they could understand how different features of successful schools compare with features of unsuccessful schools and relate this to the dimensions they were evaluating.

Sampling Methods

The recommendations resulting from this study would affect hundreds of thousands of teachers, and thousands of communities throughout Pakistan. Consequently, it was important to ensure that the conclusions were representative. Sampling methods for qualitative studies, however, have been criticized by some quantitative researchers as not being statistically representative.

The evaluation team adopted a sampling method that combined elements of both quantitative and qualitative approaches. The teshil (a district of approximately 200 villages) was chosen as the unit of analysis. Growth in enrollment of girls was selected as the key variable because the researchers were interested in understanding how individual communities made decisions over time—how they identified and overcame problems or impediments to achieving such growth.

Based on data from all teshils over 5 years, the researchers selected a statistically representative sample of 16 teshils across Pakistan (4 in each province) that showed varying levels of growth in enrollment of girls. The sample was stratified by province, and the qualitative team selected specific villages within each teshil to serve as research sites. While the sample may not have been statistically representative at the village level, it was representative at the teshil level.

Research Questions

The qualitative research focused on several key questions:

- What does a successful school look like?
- How has the school changed in the last five years?
- Why did these changes occur?

Through these questions, the researchers hoped to understand the dynamic nature of what constitutes success in enrollment of girls, including factors that support or constrain such change. They also hoped to determine whether there were any indicators of whether the changes would be sustainable in the long run.

Findings and Implications

Some of the factors that parents identified as important to their definition of school success—such as boundary walls, water, and appropriate facilities—were obvious to the researchers. Other responses were more surprising.

Above and beyond the importance of having facilities that are structurally sound, the parents indicated a desire for a certain level of aesthetic refinement, such as flowers in the schoolyard and teachers who are committed to maintaining the facilities.

The parents also expressed strong views regarding the punctuality of teachers and students. Despite the fact that they came from rural villages and may have been illiterate themselves, the parents demonstrated a clear understanding of the value of education. They felt capable of judging whether their children were learning in school, and they made decisions based on those perceptions. Moreover, their evaluation of a school's success was based in part on whether the children were well behaved, both at school and at home.

These findings suggest that there are a number of relatively simple interventions that could be built into projects that might increase their chances of success.

Another surprising finding was related to the researchers' belief that, given the government's failure in the delivery of services, the system might function better if responsibility were shifted to the community. This was confirmed to a certain extent by the research. However, the research indicated that the role of the head teacher is equally crucial. If the head teacher does not want a reform to take place, it will be difficult to overcome that constraint.

The implication for project design is that previous interventions may have overemphasized community and parent involvement, with too little attention paid to the issues of school organization or the functions of the head teacher. That, in turn, presents a policy implication. If leadership is, indeed, a critical element of improving school effectiveness, it becomes important to know how to select, train, or promote qualified individuals for those positions.

The leadership ability of the head teacher affects other areas as well. For example, absenteeism of teachers may be more effectively addressed by a collegial head teacher, who can motivate and support the teachers, than by community policing.

Successful schools seemed to have a sense of shared goals in terms of accountability and performance. However, none of the communities had parent-teacher associations or formal organizations of the type that are common in Western societies, nor had they received assistance from the World Bank or other donors. These were poor communities solving their problems in their own ways.

In practice, all of the communities that were studied faced comparable limitations in terms of school resources. Therefore, whatever success was occurring appears to be independent of teacher training, books, or instructional materials.

What differentiated the schools was the community's capacity to solve its problems. For example, the effectiveness of a school that was reasonably successful at the beginning of the five-year period of the study may decline when the government transfers the head teacher to another location. The community's ability to respond to the situation was a primary factor in the school's subsequent level of success.

This scenario further underscores the importance of the partnership between the head teacher and the community. In cases where a new headmaster was unable to forge a relationship with the community, the effectiveness of the school usually suffered. This appeared to be true regardless of whether the community had an organized approach to try to influence the government's decision.

This finding indicates that there are limits to the sustainability of interventions in Pakistan, and that political decisions or rapid turnover of teachers can threaten even successful programs.

Follow-On Research

A second phase is planned as a follow-up to the initial research. The researchers intend to use a different methodology in an attempt to revalidate the findings of the first study. The objective of the second phase is to begin to institutionalize a consultation process in which the researchers return to the field, present the findings and recommendations of the research, and ask the community if it is prepared to implement the recommendations or, if not, to suggest how they should be changed. This would in turn lead to another round of reports, recommendations, and presentations to the community, and so on in a continuing process.

Recommendations Regarding Integrated Research

The process of conducting this research convinced the author—whose background was in econometrics—of the value of using both qualitative and quantitative methods. However, the costs of this study were high, both in terms of money and time. Although the benefits of the work in terms of policy insights and improved project design justified the expense, the new budget constraints at the World Bank might make it difficult to conduct a similar study today.

In light of those constraints, researchers might consider combining the outputs of qualitative and quantitative studies or implementing several phases simultaneously in order to maximize their efforts. Another option is to collaborate with other donors in order to share the costs of the research.

It is important to remember that all research must include the participation of the client, which in most cases is the host-country government. To attempt to do otherwise would be a waste of time and resources.

At the time this chapter was written Guilherme Sedlacek was a senior education economist in the South Asia Human Resources Division and Pamela Hunte was a social scientist in the South Asia Social Development Sector Division.

Health, Water Supply, and Sanitation

8
Evaluating the Impact of Water Supply Projects in Indonesia

Gillian Brown

This chapter describes integrated research methods that were utilized in the evaluation of rural water supply projects implemented by the Government of Indonesia with support from various multilateral and bilateral agencies. The author describes the research methodologies and data analysis techniques used in these studies and presents preliminary findings of Phase 1 research, including analysis of gender issues. Advantages and disadvantages of using an integrated research approach are discussed, along with recommendations for future research.

Overview of the Research

This research was conducted in three phases. Phase 1 was part of a global study of water supply projects in five countries conducted by the Water Supply and Sanitation Unit. The survey began in 1996.

As a result of the interest generated by the first study, a second phase of research was conducted in Indonesia. Phase 2, which began in November 1997, looked at the government's own water supply project (SIPAS) and at the Asian Development Bank-supported Rural Water Supply Project.

Phase 3 of the research was initiated when UNICEF expressed a strong interest in the methodology that was being used by the research team. They noted that there appeared to be a growing body of significant data and asked if the researchers would be willing to apply these methods in looking at UNICEF's water supply and sanitation projects. Phase 3 research, which is still under way, is using a more participatory approach.

Overview of the Projects

Phase 1 looked at two World Bank projects in Indonesia. The first was a multisector Village Infrastructure Project (VIP). This project awarded grants of $12,000 to villages. The village leaders could choose what infrastructure to build from a limited menu of roads, bridges, water supply, sanitation, or piers. The villagers implemented the projects themselves, with the assistance of a field engineer from the private sector. There was very little government intervention in the project. A strong emphasis was placed on transparency.

The second project studied in Phase 1 was the Water Supply and Sanitation Project for Low-Income Communities (WSSPLIC) which was a more standard government-implemented project. The Health Department, with facilitators from the private sector and nongovernmental organizations (NGOs), dealt with the community development aspects. The Public Works Department designed the systems, which were usually built by contractors hired through the Public Works Department, although the villagers were encouraged to contribute labor for digging and carrying materials. Reflecting current trends in infrastructure projects, the project required a 20 percent contribution from the community, at least four percent in cash.

Research Methodology

Two research teams studied 16 WSSPLIC villages and 15 VIP villages. They stayed for two to three days in each village and conducted both the quantitative surveys and the qualitative interviews.

The surveyors were university students who were trained by the World Bank researcher. Their status as students was a major advantage, because they could slip into the community and engage people in casual conversations in a variety of settings. This informal interaction often provided the most useful information for the qualitative assessments.

Three surveys were conducted in each village. A household questionnaire, which was developed for the global study, was distributed to 15 randomly selected respondents. A community questionnaire was used to survey either the water groups or the community leaders, usually in a large community meeting. A technical survey was conducted, which included mapping of the entire village area to show water uses and other relevant information.

Qualitative assessments included focus group discussions with the women's leaders. Because there was so little participation by women in these projects, the surveyors had to go to the leaders to find any participation at all. In Indonesia, women are represented in village meetings by the wives of the leaders; other women are rarely consulted. Although they often are not the beneficiaries themselves, the leaders' wives are the only women who would have participated to any degree.

Photographic records were extremely useful, and they are now used as a standard procedure in all follow-up surveys. The surveyors were asked to photograph everything in the village, which helped the researchers visualize the setting if they were unable to be there themselves and aided in bringing the village to life when doing the data analysis.

In recording the qualitative data, the surveyors were instructed to "bring the village alive." They were to begin by listing ten points that made that village exactly like its neighbor; for example, both were located in the hills, or near water, and so on. They then listed ten points that made the village unique: local gossip, landmarks, or other memorable features.

Data Analysis

The analysis framework included four primary indicators, a series of secondary indicators, and subgroups within the secondary indicators.

When the quantitative survey data were analyzed using a scoring framework developed in Washington, the results in some cases contradicted the researchers' qualitative observations. In particular, the ranking of scores for certain individual indicators differed significantly from how the researchers had ranked the systems in the field.

The challenge for the researchers was to combine the qualitative and the quantitative data into a framework that was comparable with other research done under the global study so that it could be analyzed and correlated.

The researchers addressed this issue by analyzing the qualitative data again using the same secondary indicators and subgroups. This time, however, some of the subgroups were modified and others were added to refine the framework.

For each subgroup, the researchers looked at all of the villages and all of the situations they had observed for that subgroup in order to divide the villages into categories. They then ranked those categories, attached scores to that ranking, and put the new scores into the framework. Secondary indicators that included more subgroups than others were weighted accordingly before the statistical analysis took place.

Prior to the statistical analysis, the researchers ranked the villages by the scores they had originally given them in the field for each of the secondary indicators. This ranking was shown to the survey teams that had been to the villages. The statistical analysis did not take place until the entire team was satisfied that they had a scoring system for each indicator that ranked the villages in a way that was consistent with their field observations.

The survey teams returned to help with the statistical analysis, which was conducted using SPSS software. The researchers were particularly interested to find that the correlations they had observed in the field could be translated into statistical figures with appropriate regression and correlation coefficients. Their deeper understanding helped the researchers to explain the results, especially some that were contrary to expectations.

Phase 1 Findings

One of the first things evident in the frequency distributions was a double peak for the VIP project in almost every indicator. The VIP villages were at the low and high ends of the scale, and the WSSPLIC villages were in the middle. This corresponded directly with what the researchers had observed.

The main finding of the studies was that a wide variation existed in the way project rules were applied and their impact on the community. The WSSPLIC project, which was implemented through the government, managed to generate much broader participation than the VIP project. This was mainly due to the community development facilitators that were contracted through NGOs. The project rule requiring the use of facilitators, however, was not consistently applied.

The VIP project, by comparison, managed to empower the communities by handing over decision-making to them entirely. The simpler bookkeeping and transparency in VIP led to greater accountability and therefore less corruption and less waste. However, only the community leaders were involved in this decision-making, thus still falling short of empowering the water users and collectors in most cases.

The next finding was highly controversial and is still being debated. The researchers found that requiring community contributions did not increase sustainability in the WSSPLIC project, because the contributions were not linked to accountability or demand.

Current infrastructure development theory holds that community contributions are an indicator of demand. However, the researchers found that people's contributions only occasionally indicated their willingness to pay (and hence their demand) for the service. More often they paid because, for example, they were told to do so by the village head, or because well sites were auctioned off to raise money. Neither of these situations represents demand.

The researchers found that, under certain circumstances, communities were capable of building very good systems. However, more flexible options were needed for postconstruction management. The WSSPLIC project gave clear rules about establishing water committees to manage the systems. Sometimes these committees were undermined by traditional leadership structures, or vice versa. VIP had no rules, and in several cases traditional leaders or water managers were successfully managing the systems postproject. The key things for sustainability were (a) that some plan for postproject management was prepared during project implementation, and (b) that beneficiaries participated in preparing the plan and were given a range of options to choose from that corresponded to their situation and needs.

Gender Analysis

Women's participation was hard to quantify because it was so rare. As a result, the researchers had to depend primarily on qualitative assessments of gender issues.

The researchers found that the participation of water users and collectors was important. First it was necessary to identify who these were because, contrary to common assumptions, they were not always women. The village with the best system in the study had no participation from women at any level. The community built the well-constructed system and made all of the decisions, but no women were involved. This presented a problem for the researchers until they discovered that in this particular village almost all of the water was collected by the men.

The qualitative data showed that an Upland Watershed Project had introduced cows to that particular village ten years ago. The women milked the cows because the men said their hands were too hard and they didn't have the patience. This venture was very successful, and as profits grew they bought more cows, which created more work for the women. The women told the men that they could not do the milking and get water as well. Because the cows were so profitable, the men became the water collectors in that village. Hence, this was the only village surveyed where the water users and collectors were also the primary decision-makers during project implementation. This case, where the exception proves the rule, is extremely important to demonstrate the need for careful gender analysis.

Another problem was the contradiction between women's participation and empowerment of community leaders. Although the research clearly showed the advantages of handing over accountability to community leaders, in the villages where this occurred there were no women involved. Where women had participated, no accountability had been handed to the leaders. These two important factors thus nullified each other except in two cases. The first was the case recounted above, and the second was in a village with an extraordinarily powerful women's organization. The water supply systems in both villages were successful and sustainable.

Nearly all of the groups of women that were interviewed indicated they would have liked to participate more in the project, but they were never given the opportunity. However, some were not interested. For example, women who worked long and hard as tea pickers often did not want the added responsibility. This again highlights the importance of gender analysis in such projects and the danger of focusing on women per se without taking a more in-depth look at the situation.

Advantages of Integrated Research

The researchers believe there were a number of advantages to using both qualitative and quantitative methods. The first advantage became apparent in the data analysis process. The qualitative assessments informed the revision of the data analysis framework, while the quantitative analysis provided a logical explanation for those observations in a form that could be communicated effectively to others. This strengthened the researchers' confidence in the results and helped them to understand why they turned out as they did.

Most important, the analysis provided a quantified correlation between participation and both sustainability and effectiveness. This was extremely valuable because the researchers were trying to persuade the Indonesian government of the value of using more participatory, bottom-up approaches. To accomplish that objective, they needed to translate what they had observed in the field into a language that would be convincing for this audience, which would be more readily convinced by quantitative data that could be plotted on a graph.

Another major advantage of the research was the training it provided for the surveyors, who were students from the planning department of a key university in Indonesia. The selection of planning students was a strategic decision, based on the fact that a high percentage of the ministers in the current and previous government were planners who had attended that university. Some of these students may become leaders in the future, and the researchers hope they will remember what they learned through their work on this project.

Disadvantages of Integrated Research

One disadvantage of the research was that some of the findings were not statistically significant. This shortcoming is being addressed by the subsequent phases of research. The cost and the time that were required, especially for data analysis, were also cited as possible obstacles to conducting such studies in the future.

Recommendations for Future Research

It has often been said that qualitative and quantitative research should be done by separate teams. However, the researchers found there were benefits to having the same surveyors collect both the qualitative and the quantitative data for this project.

The research community needs to learn that it is important to respect both the qualitative and the quantitative approach. Arguing over which is better does not move things forward. Researchers need to say, "Let's put them together and see where it takes us," in order to make any progress. The data obtained from each approach can be put into forms that both groups can understand and use.

The author is a gender specialist for the East Asia Region of the World Bank.

9
Social Assessment of the Uzbekistan Water Supply, Sanitation, and Health Project[1]

Ayse Kudat

This chapter describes a social assessment (SA) process that was initiated in Uzbekistan. The SA's focus on social development, participation, and institutional issues required empirical research using a combination of qualitative and quantitative methodologies. The author presents the objectives of this research and the range of methodologies that were used, including the decision to establish a local social science network. The findings of the impact of the various studies are also discussed.

Background

The social assessment was launched in Uzbekistan including a series of research studies using a combination of qualitative and quantitative methodologies. The studies were undertaken by the World Bank in the context of the Uzbekistan Water Supply and Sanitation Project, which was targeted for the Aral Sea ecological disaster region.

The Aral Sea Basin encompasses an area of 690,000 square kilometers. The principal neighboring countries are Kazakstan, Tajikistan, Kyrgyzstan, and Turkmenistan. The Aral Sea lies between Kazakstan and Uzbekistan and Kazakstan in the Kyzyl Kum and the Kara Kum Deserts. In 1960, the Aral Sea was the fourth largest inland lake in the world. During the last 35 years, the sea has steadily receded as a result of the nearly total cutoff of inflow from the Amu Darya and Syr Darya Rivers due, among other factors, to heavy withdrawals for irrigation.

Today, the Aral Sea Basin is less than half of its original size and is experiencing one of the world's most serious environmental crises. In 1993, the World Bank was asked to assist in responding to some of the human needs that are emerging as a result of this crisis. Based on an initial needs assessment study, the Bank team decided that a water supply project would be an appropriate intervention.

The Research and Consultation Elements of Social Assessment

One of the earliest activities undertaken for the project was a series of 11 social studies. The social assessment process was initially part of the project identification mission; however, it became an integral part of all phases of the project cycle, including monitoring and assessment of the results of various pilot projects that were implemented as a result of the first mission.

The fact that social assessment became an integral part of so many stages of this project is attributable directly to the task manager, Roger Batstone, who strongly believed in the importance of iterative consultations and participatory fact finding throughout the process of project design and pilot implementations. It was his appreciation of the merit of social assessment that allowed the social scientists to do this work (Box 9.1).

The social assessment typically focused on a number of objectives. The first was to ensure project success and sustainability through the involvement of the stakeholders, including those it is intended to serve and those who are indirectly affected by it.

Box 9.1 Objectives of the Social Assessment

(1) Ensuring project success and sustainability through:
   The involvement of those it intends to serve in identifying their interests, needs, and priorities
   Recognition, acceptance, and responsiveness to expressed needs

(2) Assessing how water supply, sanitation, and health fit into the broader needs as perceived by the population

(3) Gaining information on:
   Household water use, sanitation, health and hygiene conditions and requirements
   Condition of household assets
   Consumption of basic food products and other necessities
   Availability of community services

A second objective was to determine how best to combine water supply, sanitation, health, and any other needs that might be appropriate. Finally, and, most important, the social assessment was launched in order to understand the socioeconomic structure, because it was the first time the Bank and its social scientists had been involved in this region.

Research Methodologies

A wide range of tools was used for the various social assessments conducted for this project, including historical analysis, secondary social science and statistical data analysis, rapid assessment, focus groups, and a household survey (Box 9.2). In selecting a particular approach, the team reviewed all of the options and considered factors such as how the study would be conducted and who would use the data. Because there was so little knowledge available to the researchers at the outset, it was determined that the first step should be an historical analysis. This enabled an understanding of key institutional issues, the dynamics of intergroup and interethnic relationships, and the nature of transformations that rural institutions and communities had experienced in the recent past.

Box 9.2 Social Assessment Tools

(1) Historical analysis

(2) Secondary data (in some case this will be sufficient to cover social assessment needs)
   - social science
   - statistical

(3) Rapid assessment (qualitative) for broad regions with high variability
   - remote sensing
   - household surveys
   - case studies
   - focus groups
   - semi-structured group interviews
   - unstructured individual discussions
   - key informant interviews
   - snowball and network techniques
   - participant observation
   - aerial photography.

The history of the Karakalpak and Uzbek relations was of great significance in terms of assessing the social risks associated with the project. The proposed project area incorporated two subregions: Karakalpakstan and Khorezm. Karakalpakstan is an autonomous republic. Its people are a minority ethnic group with a different culture and language from the population of Khorezm, who are primarily Uzbeks. Karakalpakstan is adjacent to the Aral Sea and is therefore immediately impacted by the disaster, whereas Khorezm is less directly affected. It was important to know what the response would be if project benefits were extended to those who are farther away, less impacted by the disaster, and less privileged from other perspectives.

Establishing a Social Science Network

To obtain credible information and to ensure the participation of local nongovernmental organizations (NGOs), respected institutions, and local social scientists, the researchers first established a social science network in Uzbekistan—the first ever established by the Bank. Through the network, they invited four major institutions to participate in the social assessments:

- The National Academy of Sciences, whose deputy president, an historian, was asked to lead the network
- A private marketing firm founded by a social scientist from Karakalpakstan and others who left the Academy after the opening of the free-market economy
- The Karakalpak Academy of Sciences, represented by its vice president, who is an historian and the most respected person of the region (aksakal)
- An NGO, Save the Aral Sea, which had been active in all of the international fora dealing with the Aral Sea issue.

The assignment of conducting the historical assessment and analysis was given to the Karakalpak Academy of Sciences through the network. The other institutions in the network were asked to assist the researchers in developing an overview of the demographics, socioeconomic structure, and infrastructure of the proposed project area, based on whatever statistics were available.

Needs Assessment

In addition to the historical analysis, the researchers also began to conduct a needs assessment of the thousands of village communities in the proposed project area. The purpose of this assessment was to determine if the type of project that was proposed actually corresponded to local priorities as expressed by the people themselves.

Accompanied by the Karakalpak and Uzbek social scientists, the World Bank researchers visited some 32 village communities over a three-week period. The group carried out a rapid rural appraisal, designed and pretested a questionnaire for large-scale surveys, conducted focus group discussions, and held in-depth interviews with a range of stakeholders. On the basis of those visits, the researchers realized that so little was known and the variability was so high that they could not make a judgment merely on the basis of qualitative field studies. Additional, quantitative information was required. The rural landscape was highly diverse, and there was no statistical and demographic information. As surveys were completed and systematic comparative analyses carried out, the inadequacy of results earlier achieved through rapid assessments became even more obvious. Also, the client was far more comfortable trusting results based on quantitative methods.

Household Study

To obtain additional information for the needs assessment, the researchers undertook a quantitative household survey of 951 families. The social science network—which by that time had some 40 social scientists and 40 institutions—was tapped to implement the study. Network participants in Tashkent assisted with the representative sampling strategy, while those in Karakalpakstan did the pretesting and the fieldwork.

Preliminary Findings

The findings of the initial assessments showed that the local population had three main concerns. In order of priority, these were the following:

- Lack of flour, which was scarce even when money was available
- Lack of money for food
- Lack of water (Box 9.3)

Box 9.3 Some Findings Concerning Water Supply and Distribution

Poverty is acute and widespread; people's ability to pay for services is extremely limited.

The willingness to pay for a better quality, reliable water supply is much higher than previously perceived. There is willingness to participate in financing pilot project investments through in-kind or labor contributions.

Community participation in pilots is putting pressure on local authorities to give greater attention to the scale and cost of projects. Communities welcome receiving options with respect to level of services so they can make informed decisions.

Based on these findings, it was determined that the water project should proceed, but in combination with an income-generating project. A rural credit pilot component was therefore proposed for the project.[2] Subsequent research was conducted to identify constraints to that potential component, and a strategy to overcome those constraints was formulated based on the findings of that research. To date, however, the rural credit component has not been implemented.

Subsequent Social Research

Six additional studies were conducted to finalize the SA's contributions to the project, and the results were widely disseminated. These studies included (a) a large-scale survey of urban water supply and sanitation needs and stakeholders consultation, (b) a large-scale monitoring of hand-pump water quality and community consultations on operations and management, (c) large-scale tests of salinity tolerance throughout the project areas, and (d) studies of vendors and the households that buy water from them.

Communicating the Findings

The researchers conducted a large-scale participatory workshop to discuss the findings of the initial social assessments. Two issues were actively contested: the level of income and the level of sanitation in the area, both of which the researchers had found to be extremely low. This was difficult for the government and the NGOs to accept, because they had no previous data from these regions and did not believe such conditions were possible in their country.

The data from the quantitative study, plus numerous photos and videos taken by the researchers, helped to allay some of the skepticism. To further address these concerns, the researchers took key government officials and NGO representatives to the field to observe the conditions for themselves and ask their own questions. (As a result of this experience, the NGOs made a firm commitment to support efforts to improve sanitation and income levels in these regions.)

Impact of the Social Assessments

There were considerable benefits to the use of integrated research methodologies. The quantitative data opened the door for communication with the client by convincing the government of the validity of the findings. Thereafter, the qualitative analysis moved the dialog forward.

There has been more resistance to this project from stakeholders within the Bank than in Uzbekistan. To date, the Bank has been reluctant to operationalize this initiative.

The social assessment process cost less than $200,000 to conduct, and it helped to design a $150 million project. Because of the strong support of the project task managers, the social assessment was instrumental in creating a $30 million savings to the project through one major component of the process: a salinity tolerance survey. It was also instrumental in designing many components of the project as well as in the inclusion of elements such as an urban focus that were not considered at the outset. Now, many years after the completion of this particular social assessment process, the Bank has adopted social assessment as an integral part of all projects. It is now time to revisit the Uzbekistan project and help monitor its social impacts, again using a combination of quantitative and qualitative methods.

At the time this study was conducted, the author was a social development advisor to the Europe and Central Asia Region (ECA).

Notes

1. For more information on the study, see Kudat, et al. (1997).

2. For more information on the credit component, see Europe and Central Asia Social Development Team (1996).

References

Europe and Central Asia Social Development Team. July 1996. "Social and Economic Feasibility of Rural Credit Pilot Component." Washington, D.C.: The World Bank.

Kudat, Ayse, et al. 1997. "Responding to Needs in Uzbekistan's Aral Sea Region." In *Social Assessments for Better Development: Case Studies in Russia and Central Asia*, ed. M. Cernea and A. Kudat. Environment and Sustainable Development Studies 16. Washington, D.C.: The World Bank.

10
Using Qualitative Methods to Strengthen Economic Analysis: Household Decision-Making on Malaria Prevention in Ethiopia[1]

Julian Lampietti

This study illustrates how qualitative methods can be used to enhance and complement the findings of quantitative survey methods. The author describes how qualitative methods may be used at various stages of the research process to confirm the validity of quantitative results. The objective of the research was to measure the value people place on preventing malaria in themselves and members of their household in Tigray, a province in northern Ethiopia. While it is too early to discuss the actual valuation results, it is possible to highlight the interaction between qualitative and quantitative methods.

Motivation

Malaria is one of the world's most serious infectious diseases, claiming over 2 million lives and causing 500 million cases of clinical illness every year.[2] Since their inception in the 1940s, malaria eradication efforts have met with mixed success. Resource allocation has typically been determined by centralized budgetary decisions, with little attention paid to consumer responses to the disease or to the choice of public intervention. Consumer behavior must be taken into account if the public sector is to develop financially and socially sustainable malaria prevention and treatment programs. This research takes a new, demand-side approach to measuring the value people place on preventing malaria.

Measuring the value people place on preventing malaria requires observing trade-offs between resources and the threat of illness. For example, the amount of money a person pays for chloroquine provides an estimate of what they are willing to trade off in order to prevent the disease. The challenge in this research is that there is no readily available commodity that safely eliminates the risk of malaria for an extended period of time. One solution is to ask people questions about how many hypothetical malaria vaccines they would purchase if they were available. If respondents listen to the questions and give serious answers (that is, they understand the questions and answer them honestly, taking into consideration their budget constraints), then this data can be used to estimate the value people place on preventing the disease.

There are advantages and disadvantages to asking people questions about a hypothetical vaccine. Starting with the advantages, vaccines for other illnesses are readily available, even in the most remote parts of the world. That respondents are likely to be familiar with vaccines increases one's confidence that they will take the questions seriously and make authentic choices. There may also be a positive externality to estimating the demand for a yet-to-be-developed malaria vaccine. It is widely recognized that those most likely to be afflicted with the disease are the poor. Demonstrating that even the poor are willing to pay to prevent malaria may actually increase private sector interest in vaccine development.

There are also important disadvantages to questions about hypothetical vaccines. Estimating the demand for a vaccine that does not exist may appear to have few policy implications. However, this may not be the case. Regardless of whether a vaccine is developed, knowledge of the value people place on safely and completely eliminating the risk of malaria provides information on the upper bound of what trade-offs they are willing to make in order to avoid illness. This is useful information for designing policy interventions. Another disadvantage is that describing a nonexistent malaria vaccine may unrealistically raise people's expectations and hopes. The ethical considerations of asking this type of question should be taken into account in the design of the survey instrument and research protocol.[3]

The biggest disadvantage is that a survey requires asking people what they would do rather than observing actual behavior. This is controversial because people do not have the same incentives to reveal economic commitments when faced with contingent choices as they do in real markets. In a real market, an individual's incentive to reveal the truth is that they must pay or they do not get the goods or service. However, if the subject takes the question seriously and is familiar[4] with the transaction offered in the contingent market, then he or she can probably make the same choice in a hypothetical setting as in a real market setting.

In order to minimize incentives for *strategic behavior* (e.g., respondents do not answer the question honestly in an effort to influence the outcome of the survey), respondents were asked a "yes-no" question at a randomly assigned price, followed by a question about how many hypothetical vaccines they would purchase. The yes-no format has several advantages over simply asking people how much they are willing to pay for the good.

First, it places people in a situation that is similar to many private market transactions. Because vaccinations for other illness are readily available, even in the most remote parts of the world, respondents are likely to take the contingent valuation scenario seriously and make authentic choices. Second, since only a yes or no answer is required, the format poses a relatively simple decision problem for the respondent. Third, there is no reason to expect strategic behavior in properly framed referendum questions.

Rapid Ethnographic Study

In order to measure the value people place on preventing malaria it is important to understand their perceptions of the disease. The first stage of research therefore involved a rapid ethnographic study of people in Tigray. This qualitative methodology was used to establish that people in Tigray recognize what malaria is and think of it as an important health problem, which in turn increases confidence in the reliability of the answers to the quantitative questions in the second stage of the research.

The study selected three areas in Tigray known to have malaria and was conducted by Jenkins and others (1996).[5] The field research took place in October 1996 and consisted of semistructured interviews with priests, local government officials, community health workers, and villagers. Interviews were conducted in the local language and recorded on audio cassette. They were then translated into English and entered on a computer. Thematic and textual analysis was conducted with text recovery software known as Text Collector. A total of 66 interviews were completed, of which 32 were with women and 34 with men.

The study found that respondents recognized the symptoms of malaria and considered it an important health problem and that the majority of persons interviewed associated malaria with mosquitoes. Respondents did not, however, mention the existence of a parasite actually carried by a mosquito. The rapid ethnographic survey concluded that respondents know little about prevention, wish to know more, and would be willing to pay for greater and more effective prevention.

Household Survey

The second stage involved a quantitative household survey, with the objective of estimating the value that people in Tigray place on preventing malaria. This value was to be measured in monetary units.

The survey instrument had three parts. The first part asked questions about a household's current health status, knowledge of malaria, and expenditures on malaria prevention and treatment. The second part presented respondents with questions about how many hypothetical vaccines they would purchase for their household, and the third part requested information on the socioeconomic characteristics of household members.

The hypothetical vaccine questions started by explaining to the respondent that such a vaccine was not currently, and might never become, available.[6] The enumerator then described a vaccine to the respondent, available as either a pill or an injection, which would prevent the recipient from contracting malaria for one year. The questions included a detailed description of the commodity, checked respondent understanding of how it worked, and provided reminders of substitute goods (for malaria prevention) and the budget constraint. They also emphasized that a separate vaccine would need to be purchased for each member of the household in order to protect them from getting malaria for one year.

The respondent was then asked whether he or she would purchase one or more vaccines at one of five randomly assigned prices. The price currently being charged for a bednet (mosquito net) by the local Tigray Regional Malaria Control Department, Birr 40 (US$6.00), was used as the middle price for the vaccine. The lowest price was Birr 5 (US$1.00) and the highest price Birr 200 (US$32.00). If the respondent answered yes to the original choice question, he or she was asked how many vaccines would be purchased and for whom in their household they would be used (e.g., adults, teenagers, children).

In a separate split sample, respondents received a *bednet contingent valuation scenario* that was similar to the hypothetical vaccine scenario, except that respondents were actually presented with a bednet. The split-sample design was used in order to validate the hypothetical vaccine results. This was done by comparing the determinants of demand for hypothetical vaccines with those for bednets, with the expectation that the results would be similar.

The sampling was conducted in three stages. Districts were selected in the first stage; villages were selected in the second; and households were selected in the third.[7] The questionnaire was administered by a team of 18 local interviewers, all of whom were secondary school graduates. The interviewers received two and a half weeks of training in how to administer the questionnaire. Field activities took place in January 1997 and were directed by three field supervisors. The interviews, which lasted about one hour, were conducted primarily in the early evening and early morning when heads of household were present. A total of 889 field interviews were completed.[8] Forty-one respondents were not familiar with malaria and were dropped from the study. This left 569 respondents who received the hypothetical vaccine scenario and 279 the bednet scenario. Fifty-seven percent of our respondents were female. A sad legacy of the civil war is that 34 percent of female respondents were single heads of household.[9]

Validity and Reliability of Quantitative Survey Results

The value people place on preventing malaria is a function of their economic circumstances and the severity of the disease. Given the relatively small size of their holdings, the low productivity of their land, and difficulties in gaining access to inputs and technology, many households are unable to produce enough food. Famine-causing droughts have occurred approximately every tenth year throughout this century.

These problems are compounded by the incidence of malaria, which reaches its peak during the harvest each year. In economics, the traditional welfare metric is either income or expenditure. Measuring income in rural areas is difficult: respondents often understate or overstate production, or they accidentally forget to report certain crops. Furthermore, income is variable, especially in areas where people's livelihoods depend on rain-fed agriculture.

Nonetheless, responses to survey questions about crop production, livestock holdings, and off-farm income can be aggregated to provide a measure of total household income. After using local prices to convert to crop production and livestock holdings to monetary units and annualizing where appropriate, a measure of annual household income was developed. Mean household income is Birr 1,387 (US$220) and the median is Birr 1,157 (US$183). Dividing household income by household size yields an annual per capita income estimate of Birr 300 (US$47).

This is far below the 1998 World Bank estimate of annual per capita expenditure (a proxy for income) in rural areas of Birr 945 (US$150).[10] It is also less than half the widely accepted per capita income figure of Birr 630 (US$100).[11] This raises questions about the validity of the income measure. Unfortunately, provincial-level income or expenditure estimates are not available for Ethiopia, making it impossible to determine if these differences are simply explained by differences between national and provincial measures of welfare.

One solution to this problem is to turn to the responses to qualitative questions asked in the survey. Specifically, one explanation for the very low income estimates might be that households are operating at a food deficit. Respondents were asked the qualitative question, "How much of your household's food do you grow yourself?"

Figure 10.1 indicates that, on average, lower income households grow less of their own food. Inspection of the data reveals that while a few of the households that do not grow their own food are wealthy and have other occupations, most of them are subsistence farmers and appear not to be producing enough food to support themselves. These food-deficit households probably have coping mechanisms, such as participating in local "food for work" programs. The combination of qualitative and

Figure 10.1 Comparison of Qualitative and Quantitative Questions on Food Production

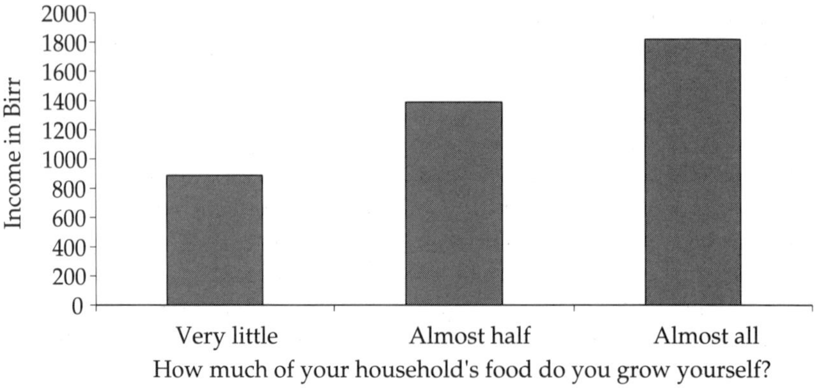

[Bar chart: Income in Birr vs. "How much of your household's food do you grow yourself?"
- Very little: ~900
- Almost half: ~1400
- Almost all: ~1800]

quantitative analysis helps explain what was not captured in the quantitative questions.

Another way to evaluate the validity of the income measure is to compare it with responses to the qualitative question, "How would you classify the economic status of your household relative to others in this village?" Table 10.1 summarizes the joint distribution of the qualitative and quantitative indicators of economic welfare. In order to make the two comparable, the quantitative measure is divided into five categories in such a way that the number of respondents in each category is equal to the number of respondents in the corresponding qualitative category. If there were total agreement between these two measures of welfare, then all of the responses would be in the diagonal cells, and the number of respondents in the nondiagonal cells would be zero.

Table 10.1 Comparison of Qualitative and Quantitative Welfare Rankings

Qualitative rank	Quantitative rank of income categories				
	1 (low)	2	3	4	5 (high)
Much worse than average	24	36	7	1	0
Below average	35	236	89	0	0
Average	8	80	233	33	4
Better than average	1	8	24	14	3
Much better than average	0	2	2	3	1

While the matching of qualitative and quantitative rankings is not perfect, it is surprisingly robust. First, the majority of responses are clustered around the diagonal cells running from left to right and top to bottom, indicating that the two measures are generally consistent. Second, both measures produce identical rankings approximately 60 percent of the time. A chi-square test of the association between the two variables is highly significant ($\chi^2_{(16)}$= 370.9). The fact that the qualitative and quantitative measures of welfare produce similar rankings increases confidence in the reliability of the quantitative income measure.

Gender differences

In the process of analyzing the data, it also became apparent that there were significant gender-based differences in agricultural income, with men reporting larger figures than women. This suggests either a systematic difference in households in which men and women were interviewed (nonequivalence of samples) or gender-based recall error (measurement error). Separating households by respondent gender and marital status reveal that female-headed households are significantly worse off than all others. This may be because of the absence of adult males to undertake plowing at the onset of the rainy season.

There is also a difference in the income reported by married men and women in households with both spouses present. Examination of nonagricultural measures of wealth in these households, such as off-farm income, housing characteristics, and household assets (lanterns, beds, radios, shoes, and jerricans) reveals that there are no significant differences between these groups. This reduces the possibility of nonequivalence of these households.

An alternative explanation can be found in the traditional division of household responsibilities. In northern Ethiopia, while both sexes are equally involved in agricultural production, men are responsible for marketing agricultural surplus (livestock and grain) and women for minding the granary. This would explain why men reported higher agricultural income than women did: they recall production while women recall consumption.

Qualitative and Quantitative Perceptions of Malaria

Malaria is endemic in the study area, with peak transmission occurring at the end of the rainy season. Because statistics on malaria incidence are not available, the study had to rely on self-reporting.

The incidence of malaria in the sample is widespread, with 78 percent of respondents reporting having had it at least once in their lifetime.

Incidence is evenly distributed across household members. In the last two years, 58 percent of respondents had malaria at least once. Fifty-three percent report that at least one other adult in their household had it, and 49 percent report that at least one teenager or child in their household had it.

The value people place on preventing malaria is in part related to their perceptions of the severity of the disease and who in the household gets sick. In order to gain better insights into this issue, respondents were asked the qualitative question, "Is malaria more serious for adults or children?" Their responses are summarized in Table 10.2. Forty-eight percent of men believe that malaria is more serious for children, while 48 percent believe that it is equally serious for adults and children. Thirty-seven percent of women believe that malaria is more serious for children, and 59 percent believe that it is equally serious for adults and children. The finding that male respondents perceive malaria to be more serious for children than female respondents counters the conventional wisdom that females are more sensitive to the health of children.

Responses to Valuation Questions

The principal choice question in the contingent valuation scenarios asked respondents how many vaccines (Table 10.3) or bednets (results not shown) they would purchase at one of five randomly assigned prices. Thirty-nine percent of respondents agreed to purchase one or more vaccines. The mean quantity purchased was four and the median three. Sixty-two percent of respondents agreed to purchase one or more bednets, and both the mean and median quantity purchased were one.

For both goods, cross-tabulations reveal the quantity purchased decreases with an increase in price.[12] This confirms that respondents listened to the price information in the scenario, and that their responses depended upon the price they received, increasing confidence in the validity of the responses to the contingent valuation questions.

Table 10.2 Perception of Seriousness of Malaria

	Number	Male	Female
Adults	34	4%	4%
Children	351	48%	37%*
Equally serious for adults and children	459	48%	59%*
Not serious for adults and children	2	0%	0%

Note: Difference is significant at 5 percent level

Table 10.3 Number of Hypothetical Vaccines Purchased by Price (n=569)

Price (Birr)	0 vaccines	1–3 vaccines	4–6 vaccines	> 7 vaccines
5	24%	35%	33%	8%
20	48%	25%	22%	6%
40	68%	21%	10%	2%
100	81%	10%	6%	3%
200	90%	6%	4%	0%

A great deal of effort was spent developing contingent valuation scenarios that would be perceived by respondents as credible in preventing malaria. As part of this effort, respondents were reminded during the interview of the total cost of their choice (e.g., price times quantity was multiplied for them) and given the opportunity to reconsider. For both goods, only 4 percent of respondents reconsidered their decision. Respondents were also asked, "How confident are you about your choice?" In the case of the vaccine, 62 percent of respondents responded "very confident," and in the case of the bednet, 78 percent responded in this manner.

The reliability of respondents' answers to these types of questions depends on the degree to which they accept or reject the basic premise underlying the choice. At the end of the survey, respondents were asked whether they believed the hypothetical vaccine or bednet would really prevent malaria. The analysis of this question indicates that respondents listened to the scenario carefully and gave serious answers to the questions. In the case of the vaccine scenario, 72 percent of respondents assumed the premise was possible. As expected, those that did not believe that it was possible were significantly more likely to say "no" to the choice question ($\chi^2_{(1)}$=50.95). For the bednet scenario, 70 percent of respondents assumed the premise was possible and, again, those that did not believe that it was possible were significantly more likely to say "no" to the choice question ($\chi^2_{(1)}$=41.79).

It is possible to use the answers to the hypothetical vaccine question to compare males' and females' stated preferences for malaria prevention in their households. This is valuable information for the design of malaria prevention programs.

Consider Figure 10.2, which exploits the information in Table 10.3, but only for married males and females (excluding single heads of household) in a hypothetical 200-household village. At prices above Birr 20 (US$3.00) per vaccine, married females' demand is greater than that of married males. Below this price there are no differences in demand by gender. At prices over Birr 20 (US$3.00), revenues and population coverage can be increased by targeting married women. For example, at a price of Birr 40 (US$6.00), an additional Birr 3,000 (US$476) in revenue can be collected by targeting women, and 7 percent more people in the 200-household village receive the vaccine.[13]

A multivariate analysis, which is not presented here, reveals two important findings about the intrahousehold allocation of malaria prevention resources. First, the number of vaccines that respondents agree to purchase decreases as adults are replaced by teenagers and children in the household. This result makes sense from an economic perspective, because children and teenagers contribute less to the household financially than do adults.

Second, there are no significant differences in the rate at which male and female respondents substitute teenagers and children for adults when choosing the optimal number of vaccines for their household and deciding who should receive them. While Figure 10.2 indicates that married women are willing to pay more for vaccines than are married men, the choice of who in the household gets the vaccines is the same. This finding is generally consistent with the qualitative results presented in Table 10.3. It is also consistent with the finding in the rapid ethnographic survey that most people thought the disease was equally serious for adults and children.

Conclusions

This research demonstrates three ways in which qualitative methods can be used to enhance and complement the findings of quantitative

Figure 10.2. Malaria Vaccine Demand for 200-Household Village in Tigray, Ethiopia

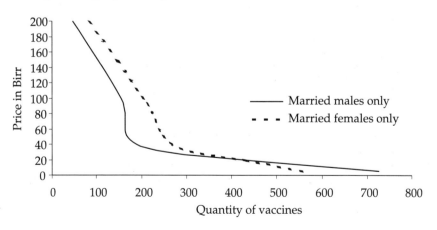

survey methods. First, a rapid ethnographic survey was used to improve the design of the quantitative survey instrument by assessing the knowledge of and attitudes towards malaria of the target population. Second, answers to qualitative questions about welfare and perceptions of malaria were combined with quantitative measures to assess validity and reliability of survey responses. This was particularly useful, for example, in assessing how seriously respondents were replying to the hypothetical questions about their willingness to purchase malaria treatments at different prices. Third, the study demonstrates that the integration of quantitative and qualitative methods, by contextualizing the quantitative findings and grounding them in a specific cultural context, provides a much broader and deeper analytical and interpretative framework than could be obtained from either of the two methods being used independently.

Julian Lampietti is a consultant with the World Bank Gender and Development Group (PRMGE).

Notes

1. The investigation received financial support from the UNDP/World Bank/World Health Organization (WHO) Special Program for Research and Training in Tropical Diseases. Gratitude is extended to C. Poulos, D. Whittington, M. Cropper, and K. Komives for participating in this research.

2. Institute of Medicine, National Academy of Sciences of the United States (1996).

3. For example, one strategy is to issue a disclaimer at the beginning of the contingent valuation scenario. Then, upon completion of data collection, debriefing meetings can be held where both the objectives and the methods used in the research are explained and subjects are given an opportunity to ask questions.

4. Through previous experiences directly paying for or exchanging something for the good or service being offered (or a similar good or service).

5. Jenkins, C. et al. 1996. "A Rapid Ethnographic Study of Malaria in Tigray, Ethiopia." Report prepared for TDR/WHO. Geneva: World Health Organization.

6. A consequence of presenting individuals with questions about a hypothetical vaccine is that it creates the expectation that the vaccine actually exists. We

tried to control both the duration and magnitude of this expectation through the study design. The questions began with a statement that research for a malaria vaccine has not been successful and that there may never be a vaccine to protect against malaria. This was followed, after the completion of data collection, with debriefing meetings in which both the objectives and the methods used in the research were explained and subjects were given an opportunity to ask questions.

7. In the first stage, two adjacent districts in Tembien sector were identified: Tangua Abergelle and Kola Tembien. In the second stage, nine villages were chosen in each of the two districts, with the assistance of local officials. They were asked to identify villages with both high and low reported treatment that could be reached from the local market towns in less than three hours by a combination of driving a four-wheel-drive vehicle and walking. The 18 villages included in the study were not selected randomly. Twelve villages received the hypothetical vaccine scenario, and six villages the bednet scenario. The third stage involved choosing a sample of households in each village. After an assessment of each village, the enumerators were instructed to walk in a specified direction and to interview every other household. Each household was presented with only one scenario, and every household in a given village was presented with the same scenario. Approximately 50 interviews were conducted in each village.

8. One record was deleted because the contingent valuation question was not completed properly.

9. A female head of household is defined as a female respondent who answered "no" to the survey question "Are you married?" or to the survey question "Is your spouse alive?"

10. As reported by the 1994/1995 rural household survey by Oxford University and Addis Ababa University. See *Ethiopia Social Sector Report*. 1998. A World Bank Country Study. Washington, D.C.: World Bank.

11. This figure is based on national income accounts, as reported by the World Bank. *World Development Report 1997*. New York: Oxford University Press.

12. The null hypothesis that the number purchased does not systematically vary with price is rejected for both goods. For hypothetical vaccines $\chi^2_{(12)} = 136.33$ and for bednets $\chi^2_{(12)} = 52.04$.

13. Of course, this would only work if the cost of targeting women over men were less than the extra revenue generated.

Women and Children

11
UNICEF'S Use of Multiple Methodologies: An Operational Context[1]

Mahesh Patel

Evaluation methodologies are selected according to the problem itself, the management decision faced, and the follow-through required. Complex problems may require use of multiple methodologies, some quantitative, some qualitative. A range of informational inputs and, hence, of evaluative methods may be needed for operational decisions and policy formulation. Required information can range from statistically valid survey data to opinions of leading politicians as obtained by key informant interviews. The most cost-effective evaluation strategy may be to optimally mix qualitative and quantitative methods, rather than use only one in a pure form.

Introduction

"There is a trade-off between the number of cases and the number of attributes of cases a researcher can study. In short, it is difficult to study social phenomena both extensively and intensively at the same time."[2]

In most monitoring and evaluation research work at UNICEF, the issue is not whether qualitative and quantitative methods should be integrated, or indeed whether they should be separated. Rather, we start with the programmatic decisions that need to be made, determine what information is needed to make those decisions, and then work out the best way to obtain that information. Different methods are used for different purposes. Sometimes, several different methods may be used together. There is nothing radical or unusual about this. Integration of different methodologies, including qualitative and quantitative methods, is routine.

This paper will describe some of the programmatic decisions, challenges, and obligations we face and the information collection methods we use to satisfy those needs. It has been structured to distinguish information collection for three different purposes: monitoring, evaluation, and research.

Monitoring

Methodologies used for monitoring UNICEF programs are often mixed, though there is a general dominance of quantitative methods for this purpose, at least at the level of national reporting. For national reporting purposes regarding achievement of global goals, such as the end-decade and mid-decade goals of the World Summit for Children, the (quantitative) household cluster survey is the normal tool, supplemented in countries with more developed data by routine information-collection systems. While data collection is based almost exclusively on counts and rates, data reporting may sometimes transform that information to rankings. In its annual report, UNICEF has ranked countries by the mortality rate of children under five years of age. The *UNDP Human Development Report* attracted considerable attention with its ranking of countries according to a composite "human development index."

Another relatively pure use of quantitative data is eradication reporting. This concerns diseases that UNICEF, with partner governments, has scheduled for eradication or elimination. Once the number of cases of a disease becomes low, the household survey is no longer a useful quantification tool. For example, the number of polio cases in many countries in southern Africa is now in the single figures. Because polio is contagious, eradication reporting requires that every case be located and identified so that immunization activity can be targeted to the actual or potential outbreak. Little additional contextual or qualitative information is required. By contrast, Guinea worm (*dracunculiasis*) eradication monitoring is somewhat different, because a community surveillance approach is being used. This requires a certain amount of contextual information regarding the community itself.

At a more general micro level, community surveillance of program progress is commonly implemented through written communications such as project and mission reports. These may contain a certain amount of quantitative information, but the dominant emphasis is on perceptions and descriptions.

It is not uncommon for UNICEF to use quantitative and qualitative methods sequentially. One example would be a quantitative household cluster survey to assess immunization coverage, followed by a qualitative assessment of the functioning of vaccine delivery systems in low-coverage areas to find out why coverage was low.[3] An assessment of cold-chain functioning might include a number of key informant interviews with health care workers at various positions in the system, supplemented by unobtrusive observation of the condition of facilities such as refrigeration equipment. Typically, the results of both the quantitative and the qualitative investigations would then be discussed in a review meeting attended by a wide range of stakeholders.

Often, different methods of analysis would be used at different levels in the monitoring process. The health status of beneficiaries might be assessed at the household level, primarily through representative quantitative sampling methods and fixed-format, or "closed," questionnaires. At a higher level of aggregation, researchers might select a few districts on a purposive basis (for example, two low-performing districts, two high-performing districts, one urban district, and one remote district) and conduct more open-ended, key informant interviews with officials in these districts to obtain their insights regarding specific problems and their ideas as to how program performance could be improved. This mix of qualitative and quantitative methods enables the organization to obtain information on the current status of core program objectives as well as insights into why differences exist and information on how to improve outcomes.

Indicators for internal monitoring of program implementation are similarly mixed, although they are predominately quantitative. Financial monitoring is of course almost entirely quantitative. The portion that is not quantitative concerns process issues such as the implementation of audit recommendations. While the evidence behind audit recommendations is often highly quantitative, the underlying judgments are often based on a certain amount of contextual information, rather than the rote application of rules in every case. Monitoring the implementation of audit recommendations often requires a relatively qualitative and descriptive approach.

One important category of implementation monitoring indicators is the UNICEF use of project milestones. A milestone is a marker for the completion of a phase in a process. Milestones in the production of a video promoting behavioral change for adolescents, for example, might be problem identification, story-line definition, production, and dissemination. Each of these milestones could be further subdivided. Story-line definition, for example, includes audience testing for comprehension, realism, credibility, feasibility of behavioral solutions depicted, potential to bring change in behavior, and entertainment value. This kind of qualitative research is an essential precondition to achieving the desired impact.

Human rights monitoring constitutes one area in which UNICEF's activity is rapidly increasing. This normally requires a wide range of methodologies, from qualitative reviews of legislation and descriptions of legal processes, to exception reporting and quantitative descriptions of how those rights are fulfilled in the population.

Evaluation

The use of quantitative and qualitative methods is much more evenly balanced in the area of evaluation. There is no clear dominance of quantitative methods in terms of either the frequency with which they are used or the importance attached to their findings. It is in fact the norm that several different methodologies are used in the course of an evaluation.

As in the case of monitoring, several different methodologies are normally used in the course of an evaluation. In one example, an immunization coverage survey was followed by an analysis of the causes of low coverage. Monitoring and evaluation are not always clearly differentiated. A staff member visiting a project site normally does both, often without any clear distinction between these functions.

Program Evaluation

UNICEF implements many different types of programs. The primary type of methodology used to evaluate the program often reflects the nature of the program itself. Some are relatively "vertical," technically driven programs, such as the effort to achieve universal salt iodization. Salt is mostly iodized in large industrial plants, and beneficiary participation in implementing or evaluating this type of technical program is normally considered a low priority.

By contrast, beneficiary participation is fundamental to all aspects of UNICEF's community-based nutrition programs. In these programs, each infant has a growth-monitoring chart. The mother regularly weighs the child and records the result on the chart. It has been verified by repeated quantitative surveys that malnutrition continued to decrease over a period of seven years in the rapidly increasing number of districts in Tanzania that are using community-based growth monitoring. In this process, monitoring, evaluation, and program implementation are inseparable and wholly participatory. Stakeholder participation in the evaluation process is closely related to participation in the program itself and intrinsic to the success of the program.

Participatory methods are also intrinsic to most process evaluation and self-evaluation exercises undertaken by UNICEF. A typical example is the Country Program Evaluation exercise, such as the one conducted in Malawi,[4] in which the entire UNICEF program in a given country may be reviewed both internally and externally. The opinions of staff and counterparts are elicited by qualitative methods such as key informant interviews and focus group discussion. In some cases, these evaluations might be performed or facilitated by a single person or team. In other instances, different persons would concentrate on different methods. As with the policy evaluations described above, review meetings that consider the results of a range of studies using varied methodologies are often key events in this evaluation process.

Research design in the field is often necessarily opportunistic. During conflict situations, for example, people are too involved to give detailed thought to collection of baseline data. The Government of Eritrea and UNICEF's joint evaluation of the orphans reunification project in Eritrea used a range of quantitative and qualitative designs.[5] The quantitative designs included one group pretest and post-test, post-test only, and one group pretest and post-test with nonequivalent groups. Qualitative methods included key informant interviews, including self-reporting and focus group work. A supplementary document review added a number of cost dimensions. The combined results showed a decrease in cost by two-thirds and better outcomes for the orphaned children as compared to the alternative strategy of supporting orphanages.

Policy Evaluation

The ultimate purpose of evaluation is to improve programs and social policy. There are many ways in which findings can be utilized. A method that is of increasing importance in this respect is the qualitative positioning of evaluation findings in the context of human rights conventions to which that country is a party. This process has been extremely effective in relation to policies in several countries concerning the incarceration of children in prison, for example.

UNICEF is also deeply involved with partner countries regarding policy reform in several other key program areas, notably in the health and education sectors. The critical event leading up to a change of policy is often a large-scale review meeting that may last several days, during which participants from different ministries and agencies present the results of their inquiries and research. These results are likely to be based on a variety of methodologies, ranging from household survey and financial implementation reports, to reports from working groups and task forces, to reviews of the recommendations of previous evaluations (which themselves may have used multiple methods, including key informant interviews, focus group discussions, mission reports, and document reviews). It is not easy to categorize this rich diversity of inputs into simple schema or to quantify the extent to which each type of methodology has contributed to reaching a decision. But it is clear that national policy-level decisions are not made on the basis of any single methodology, and that policymakers do consider diversity of input and methodology to be of significant value.

Research

As with monitoring and evaluation, the choice of research methodology primarily depends on the type of question being asked. In recent years, questions have frequently been posed concerning the impact of behavioral change messages. This area has increased in importance with the rise of the HIV/AIDS epidemic, but it is clear that behavioral change could also have a significant impact on a wide range of extremely important health issues, including major killers such as malaria and diarrheal disease. In addition, it could be a useful tool in other sectors, such as education. The success of behavioral change programs relies especially on an insightful understanding of the reasons for behavior. This type of insight depends more often on qualitative rather than quantitative research.

A significant emphasis of UNICEF's research to date has been on assessing advocacy messages through focus group discussions. In culturally sensitive areas such as messages relating to sexual behavior, we have also found it useful to interview key decision-makers regarding their perceptions of the acceptability of the proposed message. Without their approval, message distribution may not be feasible.[6]

Mixed-methodology research has been useful when a more complete description of actual behavior is needed. For example, malaria has been relatively ignored since the failure of insecticide-spray based eradication programs in the 1970s. It is now believed that a number of more effective strategies are available, primarily but not exclusively oriented around insecticide-impregnated mosquito netting for beds. Current strategies require stakeholder participation or cooperation. The consequent interest in behavioral research has led to some significant findings, among them the fact that a surprisingly high proportion of rural populations is unaware of how malaria is contracted, including the role of mosquitoes. In order to develop effective messages for program advocacy, it is important to have a deeper understanding of traditional perceptions of causality and the motivations behind household behavior in response to fever. Research in Zanzibar on these issues used mixed methods, including focus group discussions with parents and school children, key informant interviews with health-sector personnel, and a review of survey results and routine quantitative hospital information.[7]

Conclusion

Most managers seeking information on programs, if asked whether they were interested in breadth of coverage or in depth of understanding would, without hesitation, answer, both. This brings us back to the quote with which this chapter opened: "There is a trade-off between the number of cases and the number of attributes of cases a researcher can study. In short, it is difficult to study social phenomena both extensively and intensively at the same time." Difficult, perhaps, but often necessary.

There are instances in which the use of a single methodology is sufficient. But in most cases it is simpler and more appropriate to assign different methodologies to different aspects of the problem. These different methodologies may sometimes be implemented in parallel, but they are more often implemented sequentially. The results obtained using different methodologies are considered by UNICEF to be complementary and are usually discussed in a review meeting, which could be considered a mixed-method evaluative event in its own right. While there is by definition a trade-off between the number of cases and the number of attributes studied, the optimal mix of methodologies is rarely at an extreme end of the quantitative/qualitative spectrum. More commonly, the most cost-effective mix is somewhere in the middle.

In summary, the primary function of quantitative data is usually to provide information on the status of core process, outcome, and impact indicators. The primary function of qualitative reporting is to provide insight into how things are happening, why they are happening in a certain way, and what can be done to improve the situation. In some cases, one type of information is sufficient. But in many of the operational contexts described, having only qualitative or only quantitative information would restrict UNICEF's ability to attain the ultimate objective of evaluation research: the improvement of programs.

The author is the regional monitoring and evaluation officer for the UNICEF Eastern and Southern Africa Regional Office.

Notes

1. The opinions and views presented in this paper are those of the author and should not necessarily be attributed to UNICEF.

2. Ragin (1994).

3. It has rightly been commented that this assessment is an evaluation and, hence, does not belong in this section as a monitoring methodology. But monitoring and evaluation are not always easily separable. I have included cold-chain assessment here as a good example of how things work in the field. Correctly specified, this is an example of quantitative monitoring followed by qualitative evaluation.

4. Clarke (1994).

5. Morah, Mebrathu, and Sebahatu (1998).

6. Wright (1998).

7. Alilio, Eversole, and Bammek (1998).

References

Alilio M. S., H. Eversole, and J. Bammek. 1998. "A KAP study of Malaria in Zanzibar: Implications for Prevention and Control." *Evaluation and Program Planning* 21(4), 409–14.

Clarke, R. 1994. "Adapting to Democratisation: An Evaluation of the UNICEF Country Programme in Malawi." UNICEF Malawi.

Morah, E., S. Mebrathu, and K. Sebahatu. 1998. "Evaluation of the Orphans Reunification Project in Eritrea." *Evaluation and Programme Planning* 21(4), 437–48.

Ragin, C. 1994. *Constructing Social Research: The Unity and Diversity of Method.* Thousand Oaks, Calif., and London, England: Pine Forge Press. Cited in Kelly, U. 1995. *Computer Aided Qualitative Data Analysis.* London: Sage Press.

Wright K. 1998. "An Assessment of the Importance of Process in the Development of Communications Materials in Uganda." *Evaluation and Program Planning* 21(4), 415–28.

Part III
Lessons Learned

12
Lessons Learned and Guidelines for the Use of Integrated Approaches

Michael Bamberger

This chapter brings together all of the lessons learned from the workshop and the papers that were presented. It summarizes the benefits obtained from integrated approaches and shows how these can be implemented at each stage of the research process. The operational implications of integrated approaches with respect to cost, timing, and coordination are discussed, and some of the major challenges in using integrated approaches are identified.

The Benefits of Integrated Research

There is an important distinction between the integration of quantitative and qualitative data collection *methods* and the use of an integrated, multidisciplinary research *approach*. In the former, which is by far the most common, the researcher maintains the conventional (quantitative or qualitative) conceptual framework of his or her research discipline, but draws on a broader range of data collection and analysis methods, including both quantitative and qualitative. In the latter case, the researcher seeks to integrate the conceptual frameworks of two or more disciplines and to develop hypotheses from each, or all, of these disciplines. This implies a much deeper commitment to integrated approaches and normally requires the creation of a research team with senior professionals from different disciplines. The latter approach potentially offers much deeper and broader understanding of the nature and impacts of development programs and projects, but has proved more difficult to implement. Both approaches offer potentially important benefits to the research.

Benefits of Integrated Methods and Approaches

- There is a strong case for the use of integrated research, because the strengths and weaknesses of quantitative and qualitative methods often complement each other (Valadez and Bamberger 1994). (Tables 12.1 and 12.2 summarize some of the frequently cited strengths and weaknesses of quantitative and qualitative methods, respectively.)

Table 12.1 Frequently Cited Strengths and Weaknesses of Quantitative Research Methods

Strengths	Weaknesses
• Study findings can be generalized to the population about which information is required. • Samples of individuals, communities, or organizations can be selected to ensure that the results will be representative of the population studied. • Structural factors that determine how inequalities (such as gender inequalities) are produced and reproduced can be analyzed. • Quantitative estimates can be obtained of the magnitude and distribution of impacts. • Quantitative estimates can be obtained of the costs or benefits of interventions. • Clear documentation can be provided regarding the content and application of the survey instruments so that other researchers can assess the validity of the findings. • Standardized approaches permit the study to be *replicated* in different areas or over time with the production of comparable findings. • It is possible to control for the effects of extraneous variables that might result in misleading interpretations of causality.[a]	• Many kinds of information are difficult to obtain (especially with regard to sensitive issues such as sexual practices or income). • Many groups are difficult to reach (for example, women, minorities, and children). • Information may be inaccurate or incomplete. • There is no information on contextual factors to help interpret the results or to explain variations in behavior between households with similar economic and demographic characteristics. • Interview methods may alienate respondents. • Studies are expensive and time consuming; results are usually not available for a long period of time. • Research methods are inflexible because the instrument cannot be modified once the study begins.

a. For example, interviews may show that women earn less than men. This might lead to the conclusion that there are gender biases in the labor market. However, if education and years of work experience are taken into account, it might be found that gender differences in wages for people with the same education and years of work experience are much smaller.

Table 12.2 Frequently Cited Strengths and Weaknesses of Qualitative Research Methods

Strengths	Weaknesses
• Flexibility in how the methods are applied and ease in adapting to changing circumstances.[a]	• The implementation of many methods is not well documented, making them difficult to validate and replicate.[e]
• Qualitative studies are considered by many to be faster and cheaper to conduct.[b]	• Subjects are often selected without the use of sampling (and without any clearly defined criteria), so it can be difficult to generalize from the results.
• It is easier to reach difficult-to-access populations such as minorities, squatters, or women (in certain cultures).	• It is difficult to attribute responses in a group interview to specific individuals.
• Research methods can be adapted to the culture of respondents. Nonwritten and nonverbal responses are possible.	• It is difficult to control whether the interviewer is imposing responses.
• Responses can be placed in a cultural and political context.[c]	• It is difficult to analyze and interpret large numbers of case studies.
• Most research methods do not impose responses.	• Some methods may in fact not be culturally appropriate.
• Information can be obtained from groups as well as individuals.[d]	

a. For example, once a structured questionnaire has been finalized, it is not possible to change the questions or the way they are applied if problems are identified. Most qualitative methods, by contrast, can be easily adapted to changing circumstances.

b. Further comparative data is required, however, because some intensive qualitative studies can last several years and can be very expensive, whereas it is possible to conduct rapid sample surveys. For example, chapter 5 reports on a sample survey covering some 150 households that was conducted in less than two weeks.

c. For example, many qualitative methods will report on where the interview took place, who was present, and whether the respondent appeared to be constrained by other people in how she or he replied.

d. The typical quantitative survey is conducted in a one-on-one situation in which the respondent (usually of a lower status than the interviewer) replies to questions designed by outsiders. The respondent has little opportunity to discuss the issues she feels are important, or even to indicate that some of the questions do not make sense to her. On the other hand, group interviews generate a synergy between the respondents, who can focus on topics of concern to them.

e. For example, reports often do not indicate how respondents were selected, who participated in group meetings, who spoke, or what procedures were used to ensure that the reported decisions reflected the view of the majority.

Specific benefits of using integrated approaches, as cited by workshop participants, include the following:

- *Consistency checks can be built in through the use of triangulation procedures that permit two or more independent estimates to be made for key variables* (such as income, use of contraceptives, opinions about projects, reasons for using or not using public services, and so on).
- *Different perspectives can be obtained.* For example, while researchers may consider income or consumption to be the key indicators of household welfare, case studies may reveal that women are more concerned about vulnerability (defined as the lack of access to social support systems in times of crises), powerlessness, or exposure to violence.
- *Analysis can be conducted on different levels.* Survey methods can provide good estimates of individual, household, and community-level welfare, but they are much less effective for analyzing social processes (social conflict, reasons for using or not using services, and so on) or for institutional analysis (how effectively health, education, credit, and other services operate, and how they are perceived by the community). There are many qualitative methods designed to analyze issues such as social process, institutional behavior, social structure, and conflict.
- *Opportunities can be provided for feedback to help interpret findings.* Survey reports frequently include references to apparent inconsistencies in findings, or to interesting differences between communities or groups which cannot be explained by the data. In most quantitative research, once the data collection phase is completed, it is not possible to return to the field to check on such questions. The greater flexibility of qualitative research means that it is often possible to return to the field to gather additional data.
- Survey researchers frequently refer to the use of qualitative methods to check on *outliers*—responses that diverge from the general patterns. In many cases the data analyst has to make an arbitrary decision as to whether a household or community that reports conditions that are significantly above or below the norm should be excluded (on the assumption that it reflects a reporting error) or the figures adjusted. Qualitative methods permit a rapid follow-up in the field to check on these cases. Chapter 8 provides an example where researchers returned to the field to obtain additional information about the only community where women were not involved in water management. This follow-up visit revealed some extremely interesting findings.
- *Benefits and implementation strategies vary according to the professional orientation of the researcher.* The perceived benefits of integrated

research depend upon the researcher's background. From the perspective of the quantitative researcher, a qualitative component will first of all help to define key research issues and refine research hypotheses. Second, the way in which questions are formulated can be adapted to the perceptions and language of respondents, which in turn can improve the reliability and validity of responses. Third, the quantitative analysis of access to and use of project services and resources can be enhanced by analyzing the social, economic, and political context within which the project takes place. Fourth, it is possible to return to the field once data analysis is completed to clarify issues, follow up with outliers, and further explore interesting or unanticipated results.
- A qualitative researcher will find different benefits to incorporating quantitative methods. First, sampling methods can be used where necessary to ensure that findings can be generalized to the total population. Second, sample selection can be coordinated with ongoing or earlier survey research so that findings from qualitative studies can be directly compared with survey findings. And, third, statistical analysis can be used to control for household characteristics and the socioeconomic conditions of different study areas, thereby eliminating alternative explanations of the observed outcomes.
- Based on the workshop and subsequent discussions with researchers in a variety of disciplines, it is the author's impression that most quantitative researchers have few, if any, reservations about incorporating qualitative methods for data collection and analysis into their research. Many qualitative researchers, on the other hand, continue to have reservations about the validity of survey research.

Operational Approaches to the Integration of Quantitative and Qualitative Methods at Each Stage of the Research Process

Integrated research can take many forms. Table 12.3 (at the end of the chapter) documents the range of quantitative and qualitative methods used in the case studies presented in this report, and Table 12.4 (at the end of the chapter) gives examples of how quantitative and qualitative methods were combined in different studies presented in this volume.

Any research study goes through a number of stages, starting with the formulation of the research questions and ending with the publication and dissemination of the research findings. An integrated research approach should define clearly how different methods will be combined at each stage of the research.

Composition of the Research Team

The effective design and application of fully integrated approaches will normally require that the research team include principal researchers from two or more disciplines. Kozel (chapter 4) emphasizes the need to allow time and opportunities for each researcher to become familiar with the discipline of the other, and to develop mutual respect and trust among the senior researchers. For development research this will often require that this respect is developed among both the international and the national research teams.

Integrated Approaches during Formulation of Research Questions

It is important to develop a conceptual framework that draws upon two or more disciplines and to ensure that the research hypotheses draw upon each of these disciplines. This will lead to broad-based exploratory analysis that draws on the literature and research approaches of each discipline. Once this is achieved, the integration of data collection and analysis methods will normally follow automatically. Focus groups, key informant interviews, participant observation, and other qualitative methods can be used to place the proposed project or intervention within the context of community economic, cultural, and social activities. These methods can also compare priorities and concerns of different sectors of the population. Qualitative methods are particularly useful in understanding the concerns of women and other weak and less vocal groups. These methods can also be helpful in understanding some of the basic analytical concepts to be used in designing the survey instruments. For example:

- How should a household be defined in this community?
- How does the community identify the most vulnerable groups and those in need of assistance? Is the concept of "poverty," as defined by international agencies, a priority concern of the community?
- How are wealth and power defined?

As discussed in chapter 1, project outcomes are affected significantly by the social, economic, political, and cultural context within which the project is implemented. One of the most important—and underutilized —contributions of qualitative research is to help understand these different contexts or processes, to evaluate which elements are likely to influence project outcomes, and to propose ways to study them. On the other hand, sampling procedures and numerical analysis can be used to assess whether the views of key informants are shared by other sectors of the community, or to assess the relative quantitative importance of different problems and issues identified in focus groups.

Integrated Approaches during Research Design

In-depth interviews and household case studies can be helpful in defining concepts and question wording for the survey instruments. Examples of questions and concepts where wording is particularly important include:

- Defining household, family, and household membership
- Defining employment
- Measuring income and expenditures
- Defining the concept of participation in projects and political activities
- Identifying potential project impacts and understanding how to measure them
- Understanding which members of the household or group can provide information on different topics

Hentschel (1999) discusses how contextual variables can be built into the research design and how they can be used to interpret findings.

Integrated Approaches during Data Collection

Statistical sampling procedures can often strengthen qualitative research by ensuring that cases and subjects are selected so as to be representative of the populations studied. These procedures can also suggest the number of subjects required in cases where judgments are to be made about "significant" differences between groups, or about project impacts on different groups.

Qualitative methods can complement surveys by providing independent estimates or reliability checks on difficult-to-collect information such as income, household composition, employment, or participation in community decision-making. One of the most important areas in which qualitative methods can contribute to project evaluation is in helping to understand the context within which a project is implemented and the implementation process itself. Chapter 1 provides a list of specific examples that are cited in this report.

Integrated Approaches during Data Analysis and Interpretation

Once the survey data collection phase has been completed, it is usually not possible to return to the field to collect further information. As a result, many research reports include some interesting findings that cannot be explained, along with a recommendation that these be addressed in future research. Many such findings consist of unanticipated outcomes (such as target groups not using project services or using them in unexpected ways) or important differences in how subsectors respond.

A frequently cited benefit of qualitative research methods is the flexibility to conduct rapid and economical follow-up fieldwork to obtain feedback on unexplained survey findings. Several presenters noted that this is particularly helpful in the analysis of *outliers*. For example, the statistical analysis might reveal that certain families have unexpectedly high incomes, or that the participation rates of certain groups are particularly low. Traditionally, the analyst must make a judgment call as to whether these outliers are simply errors in reporting or whether they reveal important differences. Several studies cited in this report used rapid follow-up studies to throw light on these questions in a way that is usually not possible. For example, the study of the Indonesian water supply projects (chapter 8) found that women were responsible for water management in most villages. However, the surveys revealed that in one community the water supply project was managed by men. It was initially believed that this might be a reporting error, but a follow-up study found that women in this community were involved in very profitable dairy farming, from which men also benefited. Consequently, the men agreed to manage the water project so that the women could pursue an income-generating activity. This explanation provided an important insight into the project.

Qualitative methods can also be used to provide stakeholder feedback on survey findings. Statistical analysis, on the other hand, can be used to test proposed interpretations of the findings of focus groups, interviews with key informants, and so on. For example, focus group interviews with women might reveal a low level of participation in community decision-making, which might be interpreted to show gender bias and male control. Before making this interpretation, however, it would be useful to compare the findings with the proportion of men involved in decision-making to ensure that women's reported participation is actually lower.

For easy reference Table12.5 (repeating Table 1.2) presents the elements of an integrated, multidisciplinary research approach.

Conducting Research on Different Levels

Research and analysis of social development issues and programs can focus on many different levels and can examine development from a number of different perspectives. These include the perspectives of individual household members or beneficiaries of projects; individual

Table 12.5 Elements of an Integrated, Multidisciplinary Research Approach

Research Team
- Include primary researchers from different disciplines. Allow time for researchers to develop an understanding and respect for each other's disciplines and work. Each should be familiar with the basic literature and current debates in the other field.
- Ensure similar linkages among national researchers.

Broadening the Conceptual Framework
- Draw on conceptual frameworks from at least two disciplines, with each being used to enrich and broaden the other.
- Ensure that hypotheses and research approaches draw equally on both disciplines. The research framework should formulate linkages between different levels of analysis.
- Ensure that concepts and methods are not taken out of context but draw on the intellectual debates and approaches within each discipline.
- Utilize behavioral models that combine economic and other quantitative modeling with in-depth understanding of the cultural context within which the study is being conducted.

Data Collection Methods and Triangulation
- Conduct exploratory analysis to assist in hypothesis development and definition of indicators.
- Select quantitative and qualitative methods designed to complement each other, and specify the complementarities and how they will be used in the fieldwork and analysis.
- Select at least two independent estimating methods for key indicators and hypotheses.
- Utilize rigorous reporting standards for qualitative data collection and analysis procedures.

Sample Selection
- Define clearly whether, and how, qualitative findings are to be generalized.
- Where generalization to a larger population is required, ensure that appropriate statistical sampling methods are used for the selection of case studies.

Data Analysis, Follow-up Field Work, and Presentation of Findings
- Conduct and present separate analyses of quantitative and qualitative findings, and then show the linkages between the findings and levels.
- Utilize systematic triangulation procedures to check on inconsistencies or differing interpretations. Follow up on differences, where necessary, with a return to the field.
- Budget resources and time for follow-up visits to the field.
- Highlight different interpretations and findings from different methods and discuss how these enrich the interpretation of the study. Different outcomes should be considered a major strength of the integrated approach rather than an annoyance.
- Present cases and qualitative material to illustrate or test quantitative findings.

families or households; communities; districts or regions; the project or program implementation process; the institutions and organizations through which services are delivered or programs managed; and the social, political, and economic context within which projects are implemented.

Although survey methods are well suited for conducting studies at the level of the individual or household, they are often less suited for conducting analysis at the level of the community, for studying the project implementation process, for institutional analysis, or for analysis of the social, economic, and political context. However, understanding these latter levels is usually important in analyzing why projects have succeeded or failed or for understanding the factors that determine the level and distribution of outcomes and impacts. Consequently, it may be useful for the research or evaluation design to include, for example:

- A survey covering a randomly selected sample of individuals or households (to measure project outcomes and benefits and to link them to the socioeconomic characteristics of the household)
- A qualitative analysis of the project implementation process (to determine if services were delivered in a user-friendly manner and were accessible to all sectors of the community)
- A qualitative analysis of the institutions and agencies managing and regulating the project to assess how their structure, procedures, and methods of operation affected project implementation and the achievement of project objectives

Modalities of Integration

Research methods can be integrated in two main ways: sequential and simultaneous.

- *Sequential:* One method is used to prepare for the use of the other (for example, qualitative methods are used to help develop survey instruments), or one method is used as a follow-up to the other (for example, qualitative methods are used to follow up on issues or queries that arose during the coding or interpretation of survey data).
- *Simultaneous:* Both methods are used in parallel. Examples include the use of triangulation to check the validity of survey responses (for example, reported household income or sexual behavior), and the use of qualitative methods to study project implementation or to evaluate the effectiveness of the implementing institutions at the same time that quantitative surveys are being conducted.

Most of the case study presenters recommended using a sequential approach, because in practice the logistics for the simultaneous use of quantitative and qualitative methods are too complicated. This does not mean that simultaneous methods should never be used, but they are likely to require more complex planning and coordination.

Operational Implications

Based on the discussions at the workshop, a number of operational implications and guidelines are proposed for utilizing integrated research approaches.

Planning Integrated Approaches of Different Phases of the Research Cycle

Integrated methods can be beneficial at all stages of the research cycle. To be effective, however, the purposes of the integrated approach and the way in which the different methods will be used must be defined before the research begins. The effectiveness of the different methods will be increased if there is a clearly defined conceptual framework for the research, in which the key hypotheses and data requirements are defined along with the kinds of information required for each. The following are some guidelines:

- Define the conceptual framework for the research, together with the key hypotheses and data requirements.
- Define the purposes of each method before the research begins.
- Clarify where integrated methods will be used as consistency checks, where one method will be used to improve the quality of another, and where different methods will have different purposes (i.e., to obtain different kinds of information).
- Define the sequence of activities, the expected duration of each, and the overall duration of the research.
- Allow sufficient time for each phase to be completed and for the information to be processed and used in the subsequent phases.

Integrating Different Research Specialists and Teams

Multimethod approaches frequently require contracting a team of researchers with different areas of specialization. In many cases it will also be necessary to contract different teams of field researchers for the quantitative and qualitative phases of the research. This often means that the research manager must contract and work with specialists from a number of areas in which she/he is less experienced. It is important to ensure that the right kinds of specialists are contracted in each field and that good professional working relationships are established between all team members. Given the different research paradigms discussed earlier, it is extremely important to allow sufficient time and space to create a feeling of professional respect among team members.

Time Implications

The use of sequential approaches is likely to increase the duration of the study, in particular during the preparatory stages and in the analysis and interpretation phase. More time will be required at the start of the research to conduct qualitative studies to understand how the target groups think about the issues being studied and the concepts they use to discuss these issues. This may involve focus groups and case studies at the level of the individual, household, organization, or community. Additional time will be required at the end of the research to follow up on important or confusing findings and questions. Some guidelines:

- Ensure that qualitative studies are conducted well in advance of the planned date for the quantitative surveys. It is essential to allow sufficient time not only to conduct exploratory qualitative studies, but also to discuss the findings with key stakeholders and to permit them to be incorporated into the formulation and design of the sample surveys.
- Ensure that sufficient time and money are budgeted to permit follow-up fieldwork once the survey results have been analyzed. It is important to plan the follow-up as an integral part of the research and not as something that will only be done if a particular problem is found during data coding and analysis.

Cost Implications

Resources (as well as time) must be budgeted to permit the kinds of qualitative studies outlined above. Resources will also be required to contract research specialists to design, supervise, and interpret these studies. It may also be necessary to budget for a series of workshops to ensure that all of the different methodologies are fully understood by all researchers and are integrated into the research design.

Defining Professional Reporting Standards

A frequent stumbling block to the full acceptance by quantitative researchers of many qualitative studies is the lack of documentation on how the studies were designed, conducted, and analyzed. At the same time, qualitative researchers frequently complain that survey researchers use narrow, externally imposed definitions of many key concepts while ignoring the context within which projects or programs operate. Consequently, an important contribution toward better understanding and laying the groundwork for consensus among quantitative and qualitative researchers would be to develop professional standards for reporting on the methodology for all components of multimethod research projects.

The *Program Evaluation Standards* developed by the Joint Committee on Standards for Educational Evaluation (1994) provide one set of guidelines that have been widely used for reporting on, and evaluating, research in education and many other fields. Some of the key questions to be addressed in reporting on both quantitative and qualitative components of multimethod studies include:

- What methods were used to define the key issues to be addressed in the research? In particular, how were key stakeholder groups consulted?
- How were the research subjects selected? Was the sample intended to be representative of the total population that was studied? If not, what were the selection criteria?
- Are the generalizations made in the research reports justified on the basis of the way the samples were selected?
- When focus groups, or other group interviewing and consultation procedures were used, what procedures were used to determine who participated in the groups and to ensure that reported agreements and recommendations actually reflect the views of all sectors of the community (and all group members) rather than just a vocal or powerful few?
- When sample surveys were used, what procedures were employed to ensure that women and other vulnerable or difficult-to-reach groups were consulted?
- What kinds of consistency checks (triangulation) were used to ascertain the validity of reported information on key variables such as income, use of public services, sexual behavior, and other factors?
- What procedures were used to ensure effective integration of qualitative and quantitative research methods at each stage of the research process? In particular, was equal weight given to quantitative and qualitative methods in the analysis and presentation of research findings?

Challenges in Using Integrated Approaches

The use of integrated research approaches is complicated by a number of practical challenges as well as theoretical and methodological issues. Quantitative and qualitative methods were developed in different disciplines, and they reflect different research traditions. These traditions include beliefs about the ways in which data should be collected, interpreted, and used that are reinforced in the course of professional training. Differences in research paradigms, when not made explicit, can further complicate cooperative efforts to use multidisciplinary approaches when researchers from different traditions work together.

There is also a lack of experience in using integrated approaches. As a result, there are often few precedents to draw upon, leading to unanticipated delays or problems in designing or implementing integrated research.

The ongoing debate about the cost of using integrated approaches is another challenge that is reflected in the studies presented in this volume. Several of the presenters at the workshop felt that their multimethod studies, while extremely useful, were too costly to permit widespread replication.[1] Others stated that multimethod studies do not have to be excessively expensive.[2] Costs are affected by a number of factors, including the involvement of foreign consultants, local costs of conducting survey research, logistical difficulties in reaching the study areas, and other variables.

There was agreement on the fact that multimethod research increases the time and complexity of the research process. Given these complexities, several presenters felt that it is practically impossible to conduct *simultaneous* quantitative and qualitative research, in which components of the research design are modified on the basis of the findings of other components. They recommended that the studies should be *sequential*, usually beginning with qualitative work to identify hypotheses and develop instruments, followed by survey research, and ending with follow-up qualitative work to help interpret the findings.

The consensus of the workshop was that while there are considerable benefits to be gained from integrated research, further work is required to design cost-effective and operationally simple ways to conduct these studies and to keep them within a realistic budget.

The author is a senior sociologist in the Gender and Development Group of the World Bank.

Table 12.3 Quantitative and Qualitative Methods Used in the Case Studies Presented in this Report

Method	Quantitative methods Application	Cited in Chapters
Sampling procedures		
Snowball sampling	• Identifying subjects with particular characteristics	3
Sampling through references from agencies	• Identifying subjects who have received certain services	3
Random sampling	• Obtaining representative sample of working women	3
	• Analyzing interhousehold support networks	5
	• Random sample of schools	6
	• Random sample of villages	8
	• Random sample of households	9
Data collection methods		
Contingent evaluation	• Comparing preferences of men and women with respect to malaria prevention treatments	10
Household surveys	• Estimating school enrollment, participation and satisfaction with schools	7
	• Estimating socioeconomic status	8
	• Analyzing water usage	8, 9
LSMS survey with added module on vulnerability	• Studying the impacts on unstable economic conditions	5
Quantitative estimates of interhousehold network and network usage	• Analyzing survival strategies of low-income households	5
Scholastic achievement tests	• Assessing school performance	6, 7
Secondary analysis of major data sets	• Analyzing income dynamics	3
	• Analyzing women's labor force participation	3

(Table continues on the following page.)

Table 12.3 *(continued)*

Method	Quantitative methods Application	Cited in Chapters
School surveys	• Estimating women's economic contribution	3
	• Conducting needs assessments for water and related services	9
	• Obtaining information on enrollment, grade, repetition, dropout, etc.	6
	• Comparing the effectiveness of private and public school systems	7
Technical survey of water supply systems	• Analyzing water usage	8

Method	Qualitative methods Application	Cited in Chapters
Sampling procedures		
Selecting mothers through trust networks	• Analyzing coping strategies	3
Selecting case studies as subsample of quantitative survey	• Analyzing school performance and beneficiary perspectives on school reform	6
Data collection and analysis		
Case studies of individuals and households	• Analyzing the impacts of poverty	4
	• Documenting interhousehold transfers and survival strategies	4
	• Understanding how people think about malaria	10
Ethnographic survey	• Analyzing communities in which schools are located	7
	• Understanding the community definition of a successful school	7
	• Developing typologies of schools	6
Focus groups	• Obtaining beneficiary perspectives on school performance	6
	• Analyzing water leaders and water management	8
	• Understanding water needs and usage	9

Historical analysis	• Understanding stakeholder involvement	9
Informal visits with mothers	• Analyzing coping strategies	3
Key informants	• Decision-making dynamics in schools	6
	• Understanding water use and hygiene	9
	• Attitudes of communities towards malaria	10
Listing similarities and differences between villages	• "Bringing villages alive"	8
Observation of gender-based violence	• Analyzing the impacts of poverty	4
Participant observation	• Analyzing the context in which schools operate	6
	• Understanding links between reform programs and classroom behavior	6
	• Assessing condition of household assets	9
Photographic surveys	• Helping researchers and analysts to visualize the setting of each village	8
	• Understanding consumption of food and other basic necessities	9
	• Assessing availability of community services	9
Postsurvey follow-up interviews	• Understanding attitudes to malaria prevention	10
Rapid assessment studies	• Understanding expressed needs for water	9
Semistructured group interviews	• Understanding expressed needs for water	9
Social mapping	• Analyzing the impacts of poverty	4
Unstructured individual interviews	• Understanding expressed needs for water	9
Wealth ranking	• Analyzing the impacts of poverty	4

Table 12.4 Examples of the Integration of Quantitative and Qualitative Methods in the Case Studies Presented in this Report

Study	Quantitative methods	Qualitative methods	Multimethod strategy
Poverty Assessment in India (chapter 4)	Modified Living Standards Measurement Survey (LSMS) incorporating additional module on vulnerability	• Participatory rural appraisal: social mapping, wealth ranking, analysis of services and programs, analysis of social capital, analysis of gender roles and gender violence • Household case studies	Qualitative methods used to analyze the context, identify key concerns of the community, understand the concept of vulnerability, assess attitudes to poverty and perceptions as to the possibility of escaping from poverty. The methods also helped design the survey instrument. Integrated sampling frame for quantitative and qualitative studies. Households included in the wealth ranking study were also included in the sample survey so as to obtain independent estimates of wealth.
Evaluating the Impacts of Decentralization and Community Participation on Educational Quality and Participation of Girls in Pakistan (chapter 7)	• Comparison of private and public educational systems • Study as part of an experimental design to deliver schooling through government and through partnerships with government • Survey of schools, teachers, and households	• Ethnographic study at the community level to understand the constraints on the school system and the participation of girls • Evaluation of the community support program.	Quantitative and qualitative methods were used to compare the community and World Bank definitions of successful schools and to understand the constraints confronting the school system. Qualitative methods were used to help interpret the findings of the statistical analysis.
Social Assessment of	• Analysis of secondary	• Historical analysis	Quantitative and qualitative methods were

the Uzbekistan Water Supply, Sanitation and Health Project (chapter 9)	data • Household surveys • Snowball and network sampling techniques • Remote sensing • Aerial photography	• Case studies • Focus groups • Semistructured group interviews • Unstructured individual discussions • Key informant interviews • Participant observation	combined to obtain independent needs assessments, to obtain quantitative estimates of water supply and other service needs and usage, and to understand the historical, economic, political, ecological, and social context within which programs must operate.

Notes

1. For example, the Pakistan educational evaluation studies (chapter 7) were considered to have been too expensive to permit wide-scale replication.

2. In Central Asia (chapter 9), it proved possible to combine large-scale sample surveys with a range of qualitative methods at an acceptable cost.

References

Joint Committee on Standards for Educational Evaluation. 1994. *The Program Evaluation Standards.* 2d ed. Thousand Oaks, Calif.: Sage Publications.

Valadez, Joseph, and Michael Bamberger. 1994. *Monitoring and Evaluating Social Programs in Developing Countries: A Handbook for Policymakers, Managers, and Researchers.* EDI Development Studies. Washington, D.C.: The World Bank.

Bibliography

Alilio M. S., H. Eversole, and J. Bammek. 1998. "A KAP Study of Malaria in Zanzibar: Implications for Prevention and Control." *Evaluation and Program Planning* 21(4), 409–14.

Bamberger, Michael, and Daniel Kaufmann. 1984. "Patterns of Income Formation and Expenditures among the Urban Poor of Cartagena." Final Report on World Bank Research Project No. 672-57, Washington, D.C.: The World Bank.

Bamberger, Michael, Daniel Kaufmann, and Eduardo Velez. 1997. "Interhousehold Transfers and Survival Strategies of Low-Income Households: Experiences from Latin America, Africa, and Asia." Poverty Reduction and Economic Management Network, Washington, D.C.: The World Bank.

Bamberger Michael, and Jerry Lebo. 1999. Gender and Transport: A Rationale for Action. PREMNOTE No. 14 Poverty Reduction and Economic Management Network. The World Bank.

Bamberger, Michael, Mark Blacken, and Abeba Taddese. 1994. Gender Issues in Participation. Environment and Socially Sustainable Development Department. The World Bank.

Barwell, Ian. *Transport and the Village*. 1996. World Bank Discussion Paper No. 344. The World Bank.

Blackden, Mark, and Elizabeth Morris-Hughes. 1993. *Paradigm Postponed: Gender and Economic Adjustment in Sub-Saharan Africa*. AFTHR Technical Note No. 13. Africa Human Resources Department.

Brewer, John, and Albert Hunter. 1989. "Multimethod Research: A Synthesis of Styles." Sage Library of Social Research 175. Thousand Oaks, Calif.: Sage Publications.

Bryceson, Deborah Fahy, and John Howe. 1992. *African Rural Households and Transport: Reducing the Burden on Women?* IHE Working Paper IP-2. International Institute for Hydraulic and Environmental Engineering. Delft, The Netherlands.

Cernea, Michael, and Ayse Kudat, eds. 1997. *Social Assessments for Better Development: Case Studies in Russia and Central Asia.* Environmentally

Sustainable Development Studies and Monographs Series 16. Washington, D.C.: The World Bank.

Chelimsky, Eleanor, and William Shadish, eds. 1997. *Evaluation for the 21st Century.* Thousand Oaks, Calif.: Sage Publications.

Chung, Kimberly. 1997. "Using Qualitative Methods to Improve the Collection and Analysis of Data from LSMS Household Surveys (draft)." In *Designing Household Survey Questionnaires for Developing Countries: Lessons from Ten Years of LSMS Experience,* ed. Margaret Grosh and Paul Glewwe. Washington, D.C.: The World Bank.

Clarke, R. 1994. "Adapting to Democratisation: An Evaluation of the UNICEF Country Programme in Malawi." UNICEF Malawi.

Cox, Donald, and Emmanuel Jimenez. 1993. "Public and Private Safety Nets: Transfers between Households." *Outreach 13: Policy Views from the Policy Research Department.* Washington, D.C.: The World Bank.

Datta, Lois-ellin. 1997. "Multimethod Evaluations: Using Case Studies Together with other Methods." In *Evaluation for the 21st Century,* ed. Eleanor Chelimsky and William Shadish,. Thousand Oaks, Calif.: Sage Publications.

Di Gropello, Emmanuela, and Rossella Cominetti, eds. 1998. *La Descentralización de la Educación y la Salud: Un Analisis Comparativo de la Experiencia Latinoamericana.* Santiago, Chile: CEPAL (Comisión Económica para America Latina y el Caribe).

Edin, Kathryn, and Laura Lein. 1997. *Making Ends Meet: How Single Mothers Survive Welfare and Low-Wage Work.* New York: Russell Sage.

Europe and Central Asia Social Development Team. July 1996. "Social and Economic Feasibility of Rural Credit Pilot Component." Washington, D.C.: The World Bank.

Fetterman, David, S. Kaftarian, and Abraham Wandersman. 1996. *Empowerment Evaluation: Knowledge and Tools for Self-Assessment and Accountability.* Thousand Oaks, Calif.: Sage Publications.

Filmer, Deon et al. 1998. *Gender Disparity in South Asia, Comparisons Between and Within Countries.* Policy Research Working Paper 1867. Washington, D.C.: The World Bank.

Fuller, Bruce, and Magdalena Rivarola. February 1998. "Nicaragua's Experiment to Decentralize Schools: Views of Parents, Teachers and Directors." Development Economics Research Group, Washington, D.C.: The World Bank.

Goldschmit-Clermont, Luisella. 1987. *Economic Evaluations of Unpaid Household Work: Africa, Asia, Latin America, and Oceania.* Geneva: International Labor Organization.

Gottfried, Heidi, ed. 1996. *Feminism and Social Change: Bridging Theory and Practice.* Urbana: University of Illinois Press.

Grosh, Margaret, and Paul Glewwe. 1997. *Designing Household Survey Questionnaires for Developing Countries: Lessons from Ten Years of LSMS Experience.* Washington, D.C.: The World Bank.

Hanson, E. Mark. 1996. "Comparative Strategies and Educational Decentralization: Key Questions and Core Issues." University of Hong Kong.

Hentschel, Jesko. 1998. *Distinguishing Between Types of Data and Methods of Collecting Them.* Policy Research Working Paper 1914. Washington, D.C.: The World Bank.

Hentschel, Jesko. 1999. "Contextuality and Data Collection Methods: A Framework and Application to Health Service Utilization." The Journal of Development Studies Vol. 35, No. 4, April 1999, pp. 64–94.

Jain, Devaki, and Nirmala Banerjee, eds. 1985. *Tyranny of the Household: Investigative Essays on Women's Work.* New Delhi: Shakti Books.

Joint Committee on Standards for Educational Evaluation. 1994. *The Program Evaluation Standards.* 2d ed. Thousand Oaks, Calif.: Sage Publications.

Kertzer, David, and Tom Fricke. 1997. "Toward an Anthropological Demography." In *Anthropological Demography: Towards a New Synthesis*, ed. D. Kertzer and T. Fricke. University of Chicago Press.

King, Elizabeth and Berk Özler. June 1998. "What's Decentralization Got To Do With Learning? The Case of Nicaragua's School Autonomy Reform." *Impact Evaluation of Education Reforms* Working Paper Series. Development Research Group, Poverty and Human Resources Division, Washington, D.C.: The World Bank.

King, Elizabeth, Laura Rawlings, Berk Özler, and Nicaragua Reform Evaluation Team. October 1996. "Nicaragua's School Autonomy Reform: A First Look." *Impact Evaluation of Education Reforms* Working Paper Series. Development Research Group, Poverty and Human Resources Division, Washington, D.C.: The World Bank.

Kozel, Valerie. June 6, 1998. "Social and Economic Determinants of Poverty in India's Poorest Regions: Qualitative and Quantitative Assessments." India Country Program, Washington, D.C.: The World Bank.

Kudat, Ayse, et al. 1997. "Responding to Needs in Uzbekistan's Aral Sea Region." In *Social Assessments for Better Development: Case Studies in Russia and Central Asia*, ed. M. Cernea and A. Kudat. 1997. Environment and Sustainable Development Studies 16. Washington, D.C.: The World Bank.

Kumar, Krishna. 1993. *Rapid Appraisal Methods*. Washington, D.C.: The World Bank.

Lampietti, Julian. 1998. "Stated Preferences and Intrahousehold Resource Allocation: Evidence from North Ethiopia." Poverty Reduction and Economic Management Network, Washington, D.C.: The World Bank.

Mangahas, Mahar. 1999. "Monitoring Philippine Poverty by Operational Social Indicators." Paper presented at the World Bank's Poverty Reduction and Economic Management Week, July 14, 1999.

Mangahas, Mahar. 1995. "Self-Rated Poverty in the Philippines: 1981–1992." *International Journal of Public Opinion Research* 7:1.

Miles, Matthew, and A. Michael Huberman. 1994. *Qualitative Data Analysis: An Expanded Sourcebook*. Thousand Oaks, Calif.: Sage Publications.

Mohammed, Patricia, and Catherine Shepherd, eds. 1988. *Gender in Caribbean Development*. Cave Hill, Barbados: University of the West Indies, Women in Development Project.

Morah, E., S. Mebrathu, and K. Sebahatu. 1998. "Evaluation of the Orphans Reunification Project in Eritrea." *Evaluation and Programme Planning* 21(4), 437–48.

Narayan, Deepa. 1997. *Participation Tool Kit.* Washington, D.C.: The World Bank.

Nicaragua Reform Evaluation Team. 1996. "Nicaragua's School Autonomy Reform: A First Look." Poverty and Human Resources Division, Policy Research Department, Washington, D.C.: The World Bank.

Obermeyer, Carla Makhlouf. 1997 "Qualitative Methods: A Key to a Better Understanding of Demographic Behavior?" In Qualitative Methods in Population: A Symposium, *Population and Development Review*, ed. C. Obermeyer, 23(4): 813–18.

Parris, Scott. 1984. *Survival Strategies and Support Networks: An Anthropological Perspective.* Washington, D.C.: The World Bank.

Patton, Michael. 1997. *Utilization Focused Evaluation.* Thousand Oaks, Calif.: Sage Publications.

Picciotto, Robert, and Ray C. Rist, eds. 1995. *Evaluating Country Development Policies and Programs: New Approaches for a New Agenda.* San Francisco: Jossey-Boss, Inc.

Pradhan, Menno, and Martin Ravallion. 1998. "Measuring Poverty Using Qualitative Perceptions of Welfare." Policy Research Working Paper No. 2011. Development Research Group, Poverty and Human Resources, Washington, D.C.: The World Bank.

Rafferty, Adrian E., ed. 1997. *Sociological Methodology* 27. Oxford: Basil Blackwell.

Ragin, C. 1994. *Constructing Social Research: The Unity and Diversity of Method.* Thousand Oaks, Calif., and London, England: Pine Forge Press.

Rao, Vijayendra. 1997. "Can Economics Mediate the Relationship Between Anthropology and Demography?" In Qualitative Methods in Population: A Symposium, *Population and Development Review*, ed. C. Obermeyer. 23(4): 813–18.

Ravallion, Martin, and Michael Lokshin. 1999. "Subjective Economic Welfare." Policy Research Working Paper No. 2106. Development

Research Group, Poverty and Human Resources, , Washington, D.C.: The World Bank.

Reinharz, Shulamit. 1992. *Feminist Methods in Social Research.* New York: Oxford University Press.

Rietbergen-McCracken, Jennifer, and Deepa Narayan, eds. 1997. *A Resource Kit for Participation and Social Assessment.* Social Policy and Resettlement Division, Environment Department, Washington, D.C.: The World Bank.

Riffault, Helene. 1991. "How Poverty Is Perceived." In *Eurobarometer: The Dynamics of Public Opinion*, ed. Karlheinz Reif and Ronald Inglehart. London: Macmillan Academic and Professional Limited.

Salmen, Lawrence F. 1989. *Listen to the People: Participant-Observer Evaluation of Development Projects.* Oxford, England: Oxford University Press.

———. 1995. *Beneficiary Assessment: An Approach Described.* Environment Department Papers 23. Washington, D.C.: The World Bank.

Sanders, James. 1997. "Cluster Evaluation." In *Evaluation for the 21st Century*, ed. Eleanor Chelimsky and William Shadish. Thousand Oaks, Calif.: Sage Publications.

Schwartz, Norbert, and Seymour Sudman, eds. 1995. *Answering Questions; Methodology for Determining Cognitive and Communicative Processes in Survey Research.* San Francisco: Jossey-Bass, Inc.

Spalter-Roth, Roberta, Beverly Burr, Lois Shaw, and Heidi Hartmann. 1995. *Welfare that Works: The Working Lives of AFDC Recipients.* Washington, D.C.: Institute for Women's Policy Research.

Sudman, Seymour and Norman M. Bradburn. 1982. *Asking Questions.* San Francisco: Jossey-Bass, Inc.

Sudman, Seymour, Norman M. Bradburn, and Norbert Schwartz. *Thinking About Answers: The Application of Cognitive Processes to Survey Methodology.* San Francisco: Jossey-Bass, Inc.

U.S. Department of Labor Women's Bureau. 1994. *Working Women Count.* Washington, D.C.: Department of Labor.

Valadez, Joseph, and Michael Bamberger. 1994. *Monitoring and Evaluating Social Programs in Developing Countries: A Handbook for Policymakers, Managers, and Researchers.* EDI Development Studies. Washington, D.C.: The World Bank.

Watkins, Kevin, et al. 1995. The *OXFAM Poverty Report*. Oxford, England: Oxfam Publishing.

World Bank. 1996. "Improving Basic Education in Pakistan: Community Participation, System Accountability and Efficiency." Population and Human Resource Division, Country Department 1, South Asia Region, Washington, D.C.

———. 1999. *Poverty Reduction and the World Bank: Progress in Fiscal 1998.* Washington, D.C.: The World Bank.

Wright K. 1998. "An Assessment of the Importance of Process in the Development of Communications Materials in Uganda." *Evaluation and Program Planning* 21(4), 415–28.